Pitman Research Notes in Mathematics Series

Submission of proposals for consideration

Suggestions for publication, in the form of outlines and representative samples, are invited by the Editorial Board for assessment. Intending authors should approach one of the main editors or another member of the Editorial Board, citing the relevant AMS subject classifications. Alternatively, outlines may be sent directly to the publisher's offices. Refereeing is by members of the board and other mathematical authorities in the topic concerned, throughout the world.

Preparation of accepted manuscripts

On acceptance of a proposal, the publisher will supply full instructions for the preparation of manuscripts in a form suitable for direct photo-lithographic reproduction. Specially printed grid sheets can be provided and a contribution is offered by the publisher towards the cost of typing. Word processor output, subject to the publisher's approval, is also acceptable.

Illustrations should be prepared by the authors, ready for direct reproduction without further improvement. The use of hand-drawn symbols should be avoided wherever possible, in order to maintain maximum clarity of the text.

The publisher will be pleased to give any guidance necessary during the preparation of a typescript, and will be happy to answer any queries.

Important note

In order to avoid later retyping, intending authors are strongly urged not to begin final preparation of a typescript before receiving the publisher's guidelines. In this way it is hoped to preserve the uniform appearance of the series.

Addison Wesley Longman Ltd
Edinburgh Gate
Harlow, Essex, CM20 2JE
UK
(Telephone (0) 1279 623623)

Titles in this series. A full list is available from the publisher on request.

Carlos Conca

Universidad de Chile

and

Gabriel N Gatica

Universidad de Concepción, Chile

(Editors)

Numerical methods in mechanics

 LONGMAN

Addison Wesley Longman Limited
Edinburgh Gate, Harlow
Essex CM20 2JE, England
and Associated Companies throughout the world.

Published in the United States of America
by Addison Wesley Longman Inc.

First published 1997

AMS Subject Classifications   (Main) 65N30, 76M15, 76M10
                               (Subsidiary) 65N38, 65R20, 73C50

ISSN 0269-3674

ISBN 0 582 31320 1

British Library Cataloguing in Publication Data

A catalogue record for this book is
available from the British Library

Printed and bound in Great Britain
by Biddles Ltd, Guildford and King's Lynn

Table of Contents

Preface

This book contains most of the invited papers and a few selected contributed papers of the *Fourth French-Latinamerican Congress on Applied Mathematics*. This event was held at the University of Concepción, Concepción, Chile, on December 11 - 15, 1995, and its main emphasis was on the presentation of new numerical techniques in fluid and solid mechanics. The Congress was sponsored by SMAI de France (Société de Mathématiques Appliquées et Industrielles de France) and by SMAI de Chile (Sociedad de Matemáticas Aplicadas e Industriales de Chile).

The members of the Scientific Committee of the Congress are

> CLAUDE CARASSO (France)
> CARLOS CONCA (Chile)
> GABRIEL GATICA (Chile)
> VIVETTE GIRAULT (France)
> SERGIO IDELSOHN (Argentina)
> JEAN CLAUDE NEDELEC (France)
> JEAN-MARIE THOMAS (France)

The depth of the mathematical contributions as well as the wide range of applications to several problems of great practical relevance in Engineering Sciences, are the main features of this book. The first feature distinguishes this volume from similar books in Engineering Sciences, whereas the second one places it as a very important book within the Numerical Analysis bibliography. A quick glance at the table of contents reveals at once these features. Indeed, the individual contributions are briefly reviewed in the following.

The contributions by Conca and Dávila, Carasso and Panasenko, Postel and Sepúlveda deal all with the application of homogenization techniques to some problems in fluid and solid mechanics. **Conca** and **Dávila** study optimal design problems for plates with variable thickness and for the analogue determined by the diffusion equation. They use homogenization methods and duality arguments to reduce the original minimization problems to finding optimal bounds for the mixture of infinitely many materials. The latter problem is analyzed for both cases and several results are obtained. **Carasso** and **Panasenko** construct a mathematical model of a catalytic converter and apply a formal homogenization procedure to develop an asymptotic analysis of it. This yields a homogenized model of the converter, which is much simpler than the initial one, and can be solved by standard numerical methods.

Postel and **Sepúlveda** study models of propagation of a fluid polymer in a porous medium. The model considers a diffusion term and a nonlinear reaction term, and it is obtained by homogenization techniques. Also, they study numerical schemes and an inverse problem to identify some parameters of the local equations.

Several papers are devoted to some mathematical and numerical aspects of finite element methods and related procedures. **Dello Russo** and **Rodríguez** analize an eigenvalue problem arising from the computation of elastoacoustic vibrations. They use displacement variables for the fluid, to obtain a symmetric eigenvalue problem after discretization by quadrilateral Raviart-Thomas elements. In this way, the spurious modes, typical of displacement formulations, are avoided. They give a mathematical analysis of this method and compare it numerically with another previously known. **Franca** provides an overview of the residual-free-bubbles method and presents some of its applications. In particular, he examines the consequences of this discrete scheme for the Timoshenko beam problem and for a heat conduction application. He further proves that some old ideas, such as selective reduced integration and mass lumping, naturally arise from the choice of this bubble space of functions. **Idelsohn** and **Oñate** consider a finite point method previously developed by Oñate and several other authors, and extend its applicability, using a quadratic weighted least squares interpolation, to the advective-convective transport equations and to the equations governing the flow of compressible fluids. They propose a residual stabilization procedure for this finite point method, and show that the stabilization of both the convective terms and the Neumann boundary condition is necessary to ensure a correct solution in all cases. **Storti, Idelsohn** and **Nigro** present a new Petrov-Galerkin formulation to solve advection-reaction-diffusion scalar equations. They extend the standard Streamline Upwind Petrov-Galerkin method to all the situations, covering the whole plane represented by the Peclet number and the dimensionless reaction number. The new scheme is based on the inclusion of an additional perturbation function and a corresponding proportionality constant, which is selected in order to verify the super-convergence feature. **Thomas** and **Trujillo** present a new methodology for elliptic problems, called mixed finite volume method, which can be viewed as a Petrov-Galerkin primal-dual mixed formulation. They provide abstract results, similar as those in the usual Brezzi's theory for mixed finite element methods, and prove the convergence of their method on unstructured meshes.

Numerical schemes for gas dynamics problems are examined in two papers. **Allaire** and **Zelmanse** consider first the Boltzmann schemes for perfect gases, which are based on a kinetic formulation of a hyperbolic system of conservation laws. Then, they extend these kinetic schemes, at least from a numerical point of view, to gas dynamics simulations of real gases and to some two-phase flow models. For the latter case, they are concerned only with the so-called homogeneous equilibrated two-phase flow model, which describes the motion of a mixture of liquid and vapor. **Mas-Gallic** presents a numerical method for the solution of the Euler system obtained from the mass and momentum conservation equations in the study of the flow of a barotropic gas. The procedure is an extension of the particle in cell method (first introduced to compute flows of incompressible inviscid fluids), and is based on a natural separation between the quantities which are purely convected by the fluid and those which have an interaction with the fluid.

Qualitative aspects of some initial-boundary value problems in fluid mechanics are also included. **Boldrini** and **Rojas-Medar** study global existence in time of strong solutions of the three dimensional nonhomogeneous Navier-Stokes equations. These equations govern the motion of a nonhomogeneous incompressible fluid, obtained, for instance, as a mixture of miscible incompressible fluids. Under certain regularity conditions on the data, and assuming further that the external force field does not decay with time, they prove global existence and provide a sequence of estimates for the solutions and their spectral approximations. **Rojas-Medar** considers the magneto-micropolar fluid equations in a time dependent domain. He proves existence and/or uniqueness of weak and strong solutions, and gives error estimates of approximated solutions. The procedure uses first a suitable change of variables to transform the initial-boundary value problem into another problem in a cylindrical domain whose sections are time independent. Next, the new initial-boundary value problem is treated using Galerkin's approximations and the Aubin-Lions Lemma.

Some more specialized papers are also part of this proceedings. **Abboud, Nedelec** and **Zhou** use boundary integral equation formulations to solve high frequency scattering problems. For the sake of simplicity, they show the case of Helmholtz's equation only, but the same methodology applies for Maxwell's equations. They present an improvement of the usual numerical scheme which allows a mesh size of order $\lambda^{1/3}$, where λ is the corresponding wavelength. **Rivara** reviews new longest-edge algorithms for the refinement and/or improvement of triangulations. These include longest-edge propagation path of a triangle, a backward longest-edge refinement algorithm, a longest-edge improvement algorithm for Delaunay triangulations, and others. **Alvarez, Rodríguez** and **Sánchez** present an efficient numerical technique to solve scalar singular perturbation problems for a class of second order ordinary differential equations. The basic idea of the new method is to splitt the diferential operator in a sequence of stiff initial value problems, which are solved using a semi-analytical quadrature formula.

Finally, we wish to thank all the contributors, whose remarkable works were responsible for the success of the Congress. In addition, we gratefully acknowledge the financial support provided by the following institutions: *Asociación de Matemáticos Aplicados Franceses y Chilenos, Comisión Nacional de Investigación Científica y Tecnológica (CONICYT de Chile), Délégation Régionale de Coopération Scientifique et Technique, Service de Coopération Scientifique et Technique de l'Ambassade de France au Chili,* and *Universidad de Concepción* (through the *Departamento de Inge-*

niería Matemática, Facultad de Ciencias Físicas y Matemáticas, and Vicerrectoría). We are also thankful to Professor Robert P. Gilbert who accepted this volume in the Pitman Research Notes in Mathematics Series, and to Longman Higher Education for the present edition. Last, but not least, it is a pleasure to express our gratitude to our colleagues H. Alder, G. Barrenechea, E. Cisternas, A. Contreras, F. Cheuquepán, E. Figueroa, H. Mennickent, R. Naveas and F. Paiva, to the staff members C. Leiva, M.E. Ribas and J. Hernández, and to the students of the career of Mathematical Engineering of the University of Concepción, for their enthusiastic committment to the organization of the Congress.

Santiago / Concepción, May 1997

Carlos Conca / Gabriel N. Gatica

List of Contributors

ABBOUD, T.
CMAPX, Ecole Polytechnique de Paris, 91128 Palaisseau Cedex, Paris, FRANCE, e-mail: abboud@cmapx.polytechnique.fr, fax: 33-1-69333011.

ALLAIRE, G.
Commissariat a l'Energie Atomique, DRN/DMT/SERMA, C.E. Saclay, 91191 Gif sur Yvette Cedex, FRANCE, e-mail: allaire@soleil.serma.cea.fr, fax: 33-1-69082381.

ALVAREZ, L.
Instituto de Cibernética, Matemática y Física, Academia de Ciencias de Cuba, Calle E no. 309 esq. 15, Vedado, Habana, CUBA, e-mail: lilliam@cidet.icmf.inf.cu, fax: 537-333373.

BOLDRINI, J.L.
Departamento de Matematica Aplicada, IMECC-UNICAMP, CP 6065, 13081-970 Campinas, SP, BRASIL, e-mail: boldrini@ime.unicamp.br, fax: 55-192-395808.

CARASSO, C.
Equipe d'Analyse Numérique, Université Jean Monnet-Saint Etienne, 23 Rue du Dr. P. Michelon, 42023 Saint Etienne Cedex 2, FRANCE, e-mail: carasso@anumsun1.univ-st-etienne.fr, fax: 33-4-77256071.

CONCA, C.
Departamento de Ingeniería Matemática, Facultad de Ciencias Físicas y Matemáticas, Universidad de Chile, Casilla 170, Correo 3, Santiago, CHILE, e-mail: cconca@dim.uchile.cl, fax: 56-2-6712799.

DAVILA, J.
Departamento de Ingeniería Matemática, Facultad de Ciencias Físicas y Matemáticas, Universidad de Chile, Casilla 170, Correo 3, Santiago, CHILE, e-mail: jdavila@dim.uchile.cl, fax: 56-2-6712799.

DELLO RUSSO, A.
Departamento de Matemática, Facultad de Ciencias Exactas, Universidad Nacional de la Plata, CC 172, 1900 La Plata, ARGENTINA.

FRANCA, L.P.
Department of Mathematics, University of Colorado at Denver, P.O. Box 173364, Denver, CO 80217-3364, USA, e-mail: lfranca@math.cudenver.edu, fax: 303-556-8550.

IDELSOHN, S.
INTEC, Universidad Nacional del Litoral, Guemes 3450, 3000 Santa Fe, ARGENTINA, e-mail: rnsergio@alpha.arcride.edu.ar, fax: 54-42-550944.

MAS-GALLIC, S.
CMAPX, Ecole Polytechnique de Paris, 91128 Palaisseau Cedex, Paris, FRANCE, e-mail: masgalli@cmapx.polytechnique.fr, fax: 33-1-69333011.
Département de Mathématiques, Université d' Evry Val d'Essonne, 91025 Evry Cedex, FRANCE, e-mail: masgalli@lami.univ-evry.fr.

NEDELEC, J.C.
CMAPX, Ecole Polytechnique de Paris, 91128 Palaisseau Cedex, Paris, FRANCE, e-mail: nedelec@cmapx.polytechnique.fr, fax: 33-1-69333011.

NIGRO, N.
INTEC, Universidad Nacional del Litoral, Guemes 3450, 3000 Santa Fe, ARGENTINA, e-mail: nnigro@galileo.unl.edu.ar, fax: 54-42-550944.

OÑATE, E.
Centro Internacional de Métodos Numéricos en Ingeniería, Módulo C1, Campus Norte, U.P.C. Gran Capitán s/n, 08034 Barcelona, ESPAÑA, fax: 34-3-4016517.

PANASENKO, G.
Equipe d'Analyse Numérique, Université Jean Monnet-Saint Etienne, 23 Rue du Dr. P. Michelon, 42023 Saint Etienne Cedex 2, FRANCE, e-mail: panasenk@anumsun1.univ-st-etienne.fr, fax: 33-4-77256071.

POSTEL, M.
Laboratoire d'Analyse Numérique, Université Paris VI, 4 Place Jussieu, 75252 Paris Cedex 05, FRANCE, e-mail: postel@ann.jussieu.fr, fax: 33-1-44277200.

RIVARA, M.-C.
Departamento de Ciencias de la Computación, Facultad de Ciencias Físicas y Matemáticas, Universidad de Chile, Casilla 2777, Santiago, CHILE, e-mail: mcrivara@dcc.uchile.cl, fax: 56-2-6895531.

RODRIGUEZ, A.
Instituto de Cibernética, Matemática y Física, Academia de Ciencias de Cuba, Calle E no. 309 esq. 15, Vedado, Habana, CUBA, fax: 537-333373.

RODRIGUEZ, R.
Departamento de Ingeniería Matemática, Facultad de Ciencias Físicas y Matemáticas, Universidad de Concepción, Casilla 4009, Concepción, CHILE, e-mail: rodolfo@pascal.cfm.udec.cl, fax: 56-41-251529.

ROJAS-MEDAR, M.
Departamento de Matematica Aplicada, IMECC-UNICAMP, CP 6065, 13081-970 Campinas, SP, BRASIL, e-mail: marko@ime.unicamp.br, fax: 55-192-395808.

SANCHEZ, I.
Instituto de Cibernética, Matemática y Física, Academia de Ciencias de Cuba, Calle E no. 309 esq. 15, Vedado, Habana, CUBA, fax: 537-333373.

SEPULVEDA, M.
Departamento de Ingeniería Matemática, Facultad de Ciencias Físicas y Matemáticas, Universidad de Concepción, Casilla 4009, Concepción, CHILE, e-mail: mauricio@pascal.cfm.udec.cl, fax: 56-41-251529.

STORTI, M.
INTEC, Universidad Nacional del Litoral, Guemes 3450, 3000 Santa Fe, ARGENTINA, e-mail: mstorti@galileo.unl.edu.ar, fax: 54-42-550944.

THOMAS, J.-M.
Laboratoire de Mathématiques Appliquées, Université de Pau, IPRA, Av. de l'Université, 64000 Pau, FRANCE, e-mail: Jean-Marie.Thomas@univ-pau.fr, fax: 33-59-923200.

TRUJILLO, D.
Laboratoire de Mathématiques Appliquées, Université de Pau, IPRA, Av. de l'Université, 64000 Pau, FRANCE, e-mail: David.Trujillo@univ-pau.fr, fax: 33-59-923200.

ZELMANSE, A.
Commissariat a l'Energie Atomique, DRN/DMT/SERMA, C.E. Saclay, 91191 Gif sur Yvette Cedex, FRANCE, fax: 33-1-69082381.

ZHOU, B.
CMAPX, Ecole Polytechnique de Paris, 91128 Palaisseau Cedex, Paris, FRANCE, e-mail: zhou@cmapx.polytechnique.fr, fax: 33-1-69333011.

TOUFIC ABBOUD, JEAN-CLAUDE NÉDÉLEC and BIN ZHOU

Recent developments on a phase separated integral equation method for high frequency scattering problems

1   Introduction

The Boundary Integral Equation Method (BIEM) is commonly used to solve diffraction problems. In the case of constant coefficients the solution of the exterior problem can be represented by integrals on the boundary only. It has been noticed that in order to have a good result the mesh size should be proportional to the wavelength λ and of order $\lambda/10$ in practice, which is a serious handicap for high frequency calculations. In this paper, we present an improvement of the method which allows a mesh size of order $\lambda^{1/3}$. For the sake of simplicity, we present the case of Helmholtz equation only, but the same analysis holds for Maxwell's equations.

We first fix some notations : Ω is a bounded open domain in $\mathbb{R}^2$ or $\mathbb{R}^3$, Γ its boundary, supposed to be smooth, $\vec{n}$ is the outward normal to Ω, Ω' is the complement of $\overline{\Omega}$. We consider the diffraction problem :

$$(P0)\begin{cases} \Delta u + k^2 u = 0 & \text{in } \Omega', \\ \dfrac{\partial u}{\partial n} + i\,k\,Z_r\,u = g & \text{on } \Gamma, \\ u \text{ satisfies the out-going wave condition.} \end{cases}$$

Here u is the diffracted field, and Z_r is the relative impedance of the boundary, with $\Re e(Z_r) \geq 0$ for physical reason. The Neumann and Dirichlet conditions are just special cases of $Z_r = 0$ and $Z_r \to \infty$, and

$$g = -\frac{\partial u^{\text{in}}}{\partial n} - i\,k\,Z_r\,u^{\text{in}}.$$

The incident wave u^{in} satisfies the free-space Helmholtz equation. The radiation (out-going) wave condition is, for $n = 2, 3$

$$(1) \qquad \frac{\partial u}{\partial r} - i\,k\,u = o(r^{(1-n)/2}) \qquad\qquad r \to \infty.$$

We will recall in the section 2 and section 3 some results on the existence and uniqueness of the solution, and establish the boundary integral equation representation for the solution of (P0), the corresponding discrete problem and the consistent error estimates will be given. In section 4, we expose the new idea, the matrix calculation and results are given in section 6 and 7. Finally, we conclude in section 8.

1

2 Boundary integral representation

By truncating the exterior domain and using the Dirichlet-Neumann operator on the truncation boundary, we can establish the existence and uniqueness of solution to (P0) (cf. [6]). However, by associating an interior problem to (P0) we can obtain a boundary integral representation, from which a direct proof is possible, and (P0) is solved in an elegant manner.

We extend solution u of (P0) to Ω as the solution of the following interior problem :

$$(P1) \begin{cases} \Delta u + k^2 u = 0 & \text{in } \Omega, \\ \dfrac{\partial u}{\partial n} + i k Z_r u = g & \text{on } \Gamma. \end{cases}$$

If $\Re(Z_r) \neq 0$ or if $-k^2$ is not an eigenvalue of the Laplacian when $\Re(Z_r) = 0$, the solution is unique (while $-u^{\text{in}}$ is always one solution). Despite the simplicity of this choice, other choices of boundary condition are possible and lead to different integral representations.

We represent the combined solution of $(P0), (P1)$ by the following formula :

$$u(x) = \int_\Gamma G(k|x-y|) \left[\frac{\partial u}{\partial n}\right](y)\, d\gamma(y) - \int_\Gamma \frac{\partial G}{\partial n}(k|x-y|)\, [u](y)\, d\gamma(y) \qquad x \in \Omega \cup \Omega',$$

(2)

where $G(r)$ is the Green's function of the Helmholtz equation, satisfying the out-going wave condition (1),

$$G(kr) = \frac{i}{4} H_0^{(1)}(kr) \qquad \text{in 2D,} \qquad\qquad G(kr) = \frac{e^{ikr}}{4\pi r} \qquad \text{in 3D,}$$

and we use the notation $[\cdot]$

$$[f](x) = \lim_{\Omega \ni y \to x \in \Gamma} f(y) \;-\; \lim_{\Omega' \ni y \to x \in \Gamma} f(y)\,.$$

Take a simplified notation : let $q = [u]$, then $[\partial u/\partial n] = -i k Z_r\, q$. The potential u defined by (2) satisfies Helmholtz equation in Ω' and the radiation condition (1). In order to determine q, we take the outside (or inside) limit of u and $\partial u/\partial n$, the boundary condition of (P0) (or (P1)) leads to the following integral equation (cf. [6]) :

(3)
$$I_1(q) + I_2(q) + I_3(q) = g,$$

where

$$I_1(q) = -\text{``}\int_\Gamma \frac{\partial^2 G}{\partial n_x \partial n_y}(k|x-y|)q(y)\, d\gamma(y)\text{''}\,,$$

$$I_2(q) = k^2 \int_\Gamma Z_r(x) Z_r(y) G(k|x-y|) q(y)\, d\gamma(y)\,,$$

$$I_3(q) = -ik \int_\Gamma Z_r(y) \frac{\partial G}{\partial n_x}(k|x-y|) q(y)\, d\gamma(y)$$

$$\qquad\quad -ik \int_\Gamma Z_r(x) \frac{\partial G}{\partial n_y}(k|x-y|) q(y)\, d\gamma(y)\,,$$

I_1 is an integral in the sense of Cauchy principal value (quoted for emphasize).

2

3 Variational formulation and Finite Element approximation

From equation (3), we have, for all $q' \in H^{1/2}(\Gamma)$,

$$A_1(q, q') + A_2(q, q') + A_3(q, q') = b(q') \,,$$

where

$$
\begin{aligned}
A_1(q, q') &= -\text{``} \int_\Gamma \int_\Gamma \frac{\partial^2 G}{\partial n_x \partial n_y}(k|x - y|) q(y) \overline{q'}(x) \, d\gamma(y) d\gamma(x) \text{ ''} \,, \\
A_2(q, q') &= \int_\Gamma \int_\Gamma k^2 Z_r(x) Z_r(y) G(k|x - y|) q(y) \overline{q'}(x) \, d\gamma(y) d\gamma(x) \,, \\
A_3(q, q') &= -ik \int_\Gamma \int_\Gamma Z_r(y) \frac{\partial G}{\partial n_x}(k|x - y|) q(y) \overline{q'}(x) \, d\gamma(y) d\gamma(x) \\
&\quad -ik \int_\Gamma \int_\Gamma Z_r(x) \frac{\partial G}{\partial n_y}(k|x - y|) q(y) \overline{q'}(x) \, d\gamma(y) d\gamma(x) \,, \\
b(q) &= \int_\Gamma g(x) \overline{q'}(x) d\gamma(x) \,.
\end{aligned}
$$

From [9] , the integral A_1 is equal to the following form, easier to calculate :

$$A_1'(q, q') = \int_\Gamma \int_\Gamma G(k|x - y|) \left(\vec{\mathrm{rot}}_\Gamma \, q(y) \cdot \vec{\mathrm{rot}}_\Gamma \, \overline{q'}(x) - k^2 \, \vec{n}(x) \cdot \vec{n}(y) \, q(y) \, \overline{q'}(x) \right) \, d\gamma(y) d\gamma(x) \,.$$

We can now take a concise notation :

$$A = A_1' + A_2 + A_3 \,.$$

The boundary integral problem is

$$(P2) \begin{cases} \text{Find } \quad q \in H^{1/2}(\Gamma) \quad \text{such that} \quad \forall q' \in H^{1/2}(\Gamma) \\ A(q, q') = b(q') \,. \end{cases}$$

The problem (P2) is well posed, we can prove

THEOREM 3.1. *The mapping $g \in H^{-1/2}(\Gamma) \longmapsto q \in H^{1/2}(\Gamma)$ is an isomorphism if $\mathrm{Mes}\{\Re e(Z_r) > 0\} > 0$.*

Remark :   This formulas gives a natural approximation to the Dirichlet and Neumann boundary condition cases when $-k^2$ is a eigenvalue of the interior problem.

Discretization

There have two approximations here : First we approximate the boundary Γ by a piecewise l-degree polynomial one Γ_h, and we denote π the orthogonal projection from Γ_h on Γ, which is a bijection if h is small enough. Then we approximate the variational spaces $H^{-1/2}(\Gamma)$ and $H^{1/2}(\Gamma)$ by spaces of piecewise polynomials of degree m, that is

$$V_0(\Gamma_h) = \left\{ q_h \in L^2(\Gamma_h) : \quad \forall \text{ element } K \in \Gamma_h, q_h|_K \text{ is constant} \right\} \qquad \text{if } m = 0 \,,$$

$$V_m(\Gamma_h) = \left\{ q_h \in C^0(\Gamma_h) : \forall \text{ element } K \in \Gamma_h, q_h|_K \text{ is a polynomial of degree } m \right\} \ m \geq 1 \,.$$

3

Denote by q_h the solution of the discrete problem

$$(P3) \begin{cases} \text{Find} \quad q_h \in V_m(\Gamma_h) \quad \text{such that} \quad \forall q' \in V_m(\Gamma_h) \\ A_h(q_h, q'_h) = b_h(q'_h) \, . \end{cases}$$

where A_h, b_h are integrals on Γ_h with in the integrand, Z_r replaced by its pull-back of π on Γ_h. We have from [8] the following error estimate :

<u>Dirichlet case</u>

$$q = \left[\frac{\partial u}{\partial n} \right] \, , \qquad\qquad g = -u^{\text{in}} \, ,$$

$$(4) \quad \|q - \pi(q_h)\|_{L^2(\Gamma)} \le C \left\{ \|g - \pi(g_h)\|_{H^{1/2}(\Gamma)} + h^{m+1} \|q\|_{H^{m+1}(\Gamma)} + h^l \|q\|_{L^2(\Gamma)} \right\} \, ,$$

We deduce with the same techniques as in [8], with $[u]$ as unknown,
<u>Neumann case</u>

$$q = [u] \, , \qquad\qquad g = -\frac{\partial u^{\text{in}}}{\partial n} \, ,$$

$$(5) \quad \|q - \pi(q_h)\|_{L^2(\Gamma)} \le C \left\{ \sqrt{h} \|g - \pi(g_h)\|_{H^{-1/2}(\Gamma)} + \|g - \pi(g_h)\|_{H^{-1}(\Gamma)} \right.$$
$$\left. + h^{m+1} \|q\|_{H^{m+1}(\Gamma)} + h^{l+1} \|q\|_{H^{1/2}(\Gamma)} \right\} \, .$$

The mixed boundary condition case is analogous because all error estimates involved are done in Dirichlet and Neumann cases.

4 Early works and the new idea

While the BIEM works in low and middle range frequencies, it is hard to apply it to high frequencies because of its mesh criterion. One advantage of this method is that the radiating wave condition is automatically satisfied. On counterpart, we have a full filled matrix. Other methods, like the exterior domain finite element or finite difference methods, are costly and sometimes inaccurate not only for the reason of mesh criterion, but also because of the difficulty to approximate the radiation condition for all boundary forms. The geometric optics (GO) method, or the "ray method", gives a rapid and good precision approximation in most regions, in particular at high frequency cases. It is therefore natural to use some information of GO to accelerate the BIEM.

The first attempt is to use the GO approximation of current q in the domains where this approximation is accurate, that is to say, in the "illuminated" and "dark" regions only. This works well but there are discontinuities at the separation point – the boundary of the regions in which we use a GO approximation, and there is less interest in 3D cases because the transition region is larger than in 2D cases (though of same order), the number of degrees of freedom becomes $O(\lambda^{1/3}/\lambda^2) = O(\lambda^{-5/3})$ in 3D whereas $O(\lambda^{-2/3})$ in 2D. The second idea is to write $q(x) = \zeta(x)e^{ik\varphi(x)}$ and to look for the modulus ζ and phase φ using a finite element method. This method, with the

4

implicit hypothesis that the modulus and phase functions don't change "dramatically", which is the same hypothesis *a posteriori* of the GO, leads to a nonlinear problem, and we don't even know if there is a solution to the corresponding discrete problem! Numerical experiments show that the least-square solution of this nonlinear problem is not satisfactory even in the simplest case. So it is not practical. Then comes the idea to look for the current q as a product of two factors : we write $q(x) = e^{ik\varphi_0(x)}\tilde{q}(x)$ with φ_0 the first degree approximation of the phase function, which is more accurate in the illuminated and dark regions than in the transition region, and this function can be calculated explicitly. $\tilde{q}$ is the new unknown. This method is somewhat a mixing of the above two tentatives. Our aim is reached : we have separated the "phase" and "modulus" part, in the same time, the problem remains linear. As we have isolated the part which most contributes to the oscillations of the current function, the second factor $\tilde{q}$ must be more moderate, and relatively easy to calculate in high frequency cases.

The idea to use sinusoidal test functions is in fact not new, other authors have also used some nonconventional test functions. F. X. Canning [5] recently used some smooth overlapping test functions with the moment-method, and localized the important reactions to only a small number of elements with the untapered directional basis functions; A. de La Bourdonnaye [10] studied the use of stationary phase method with exponential test functions; Aberegg & Peterson [3] have separately found a very analogous method for the 2D scattering problem.

This method is introduced to resolve the high frequency diffraction problems, but we see also some adaptation to time-dependent acoustic wave scattering problems (cf. [2]).

5 New finite element method and the boundary approximation

We take the finite element space

$$\tilde{V}_m(\Gamma_h) = \left\{ q(x) = e^{ik\varphi_0(x)}\tilde{q}(x) : \tilde{q} \in V_m(\Gamma_h) \right\} \ .$$

In the plane incident wave case, we have the following result, due to Melrose and Taylor [11] :*We suppose that Ω is a convex C^∞ manifold of $\mathbb{R}^n$, the incident wave is plane $u^{in} = e^{ik\vec{\nu}\cdot\vec{x}}$, then with Dirichlet condition, the exterior normal derivative of u on Γ at a vicinity of the separation point $(\vec{\nu}\cdot\vec{n} = 0)$ in the transition region has the following expansion:*

$$\frac{\partial u}{\partial n} = K(k, \vec{\nu}, \vec{x})e^{ik\vec{\nu}\cdot\vec{x}} \ ,$$

$$K(k, \vec{\nu}, \vec{x}) = k^{2/3}\frac{\vec{\nu}\cdot\vec{n}(x)}{Z}\Psi(k^{1/3}Z) + S^0_{2/3,1/3} \ ,$$

and with Neumann or Robin (mixed) boundary conditions :

$$u = K_N(k, \vec{\nu}, \vec{x})e^{ik\vec{\nu}\cdot\vec{x}} \ ,$$

5

$$K_N(k, \vec{\nu}, \vec{x}) \;=\; b(\vec{\nu}, \vec{x})\Phi(k^{1/3}Z) + S^{-1/3}_{2/3,1/3}(\Gamma, \mathbb{R}) \; ,$$

where Z and b are C^∞ functions, Z vanishes at first order at $\vec{\nu} \cdot \vec{n}(x) = 0$, Ψ and Φ are rapidly decreasing functions at $+\infty$, all these functions are independent of the wave number. Here $S^m_{\rho,\delta}$ are symbol spaces (cf. [12] or [13] for details), the asymptotic expansions of Ψ and Φ at $\pm\infty$ give a first degree approximation of $\partial u/\partial n$ and u which coincides with that of GO. We have in this case $\varphi_0 = \vec{\nu} \cdot \vec{x}$, and because for the interior solution, we have $u|_{\Gamma_-} = e^{ik\,\vec{\nu}\cdot\vec{x}}$, and $(\partial u/\partial n)|_{\Gamma_-} = ik\vec{\nu} \cdot \vec{x}\, e^{ik\,\vec{\nu}\cdot\vec{x}}$, we have the following estimate for $\tilde{q}$:

$$\|\tilde{q}\|_{H^m(\Gamma)} = O(k^{1+m/3}) \qquad m \geq 0 \; ,$$

while

$$\|q\|_{H^m(\Gamma)} = O(k^{m+1}) \qquad m \geq 0 \; ,$$

which implies with (4) and (5) (cf. [15] for details):

$$\|q - \pi(q_h)\|_{L^2(\Gamma)} \leq C \left(C_1\, k(hk)^{m+1} + E_h(k) + f_h(g) \right) \; ,$$

whereas for the new finite element

$$\|\tilde{q} - \pi(\tilde{q}_h)\|_{L^2(\Gamma)} \leq C \left(C_2\, k(hk^{1/3})^{m+1} + \tilde{E}_h(k) + \tilde{f}_h(g) \right) \; .$$

Here C is the same constant as in (4) and (5), and is generally a function of k, but C_1 and C_2 are independent of k. In the case of Dirichlet or Neumann boundary condition C is not bounded if k approaches a critical wavenumber of the interior problem (P1), but it seems to be bounded in the mixed boundary condition case.

The term $E_h(k), \tilde{E}_h(k)$ are errors from the boundary approximation, respectively in the two finite element cases. We'll come back to their estimations later. We'll see that the boundary approximation is more important in our new finite element method. The last terms $f_h(g)$ and $\tilde{f}_h(g)$ in (4) and (5) are small compared to the first terms when $k \to \infty$, so these terms are not important.

We also have by the asymptotic expansion for large k

$$\|q\|_{L^2(\Gamma)} = \|\tilde{q}\|_{L^2(\Gamma)} \geq C_0\, k \; ,$$

so the following relative error estimates hold :

$$\frac{\|q - \pi(q_h)\|_{L^2(\Gamma)}}{\|q\|_{L^2(\Gamma)}} \leq C' \left(C_1(hk)^{m+1} + k^{-1}E_h(k) + \frac{1}{k}f_h(g) \right) \; ,$$

$$\frac{\|\tilde{q} - \pi(\tilde{q}_h)\|_{L^2(\Gamma)}}{\|\tilde{q}\|_{L^2(\Gamma)}} \leq C' \left(C_2\,(hk^{1/3})^{m+1} + k^{-1}\tilde{E}_h(k) + \frac{1}{k}\tilde{f}_h(g) \right) \; .$$

The *expected* leading terms are respectively $(hk)^{m+1}$ and $(hk^{1/3})^{m+1}$; this is true for the first equation if $l \geq 1$ (we'll see the error estimate later). So we justify the empiric mesh criterion of the classical BIEM : $hk \leq cst$, which is $h \leq c\lambda$. In our new finite

6

element case, if we suppose there is no error in the boundary approximation, we must only keep $hk^{1/3} \leq cst$ to have a good numerical result. This estimate is valid for both 2D and 3D cases, so we can expect to use a smaller number of degrees of freedom, of order $k^{1/3}$ in 2D case and $k^{2/3}$ in 3D case.

When there are geometry approximations, we can see that with the increase of the wavenumber, the difference between the approximated boundary Γ_h and the real boundary Γ "grows" if no refinement is done, the error due to the geometry approximation will be more important than that of the function space. It is also a basic condition that this distance remains small to prove the existence of solution to the discrete problem. So we can never take a mash size independent of the wave lengths. On the other side, if we want to improve the phase approximation, say, including in the phase part $ik\varphi$ other terms, as the second term of the phase development in series of k for example, making the phase function like $i(k\varphi_0 + k^{1/3}\varphi_1)$ (it's in this case that we have a development of $\tilde{q}$ in series of k with no positive power number except the first term, most physicians search the Ansatz in form of a modulus and a exponential part like this, see also (10)), the function φ_1 is generally of complex form, and the integrals will be much more complicated than that we have here. Effectively we can expect a better approximation to the function space, but to obtain really a fundamental improvement, the prices to be payed are the complexity in the boundary approximation, and accurate approximation of the oscillating integrals.

We now come to control the boundary approximation error $E_h(k)$. More studies on the geometry approximation give the following estimations, in the Dirichlet and Neumann case respectively:

$$
\begin{aligned}
E_h(k) &\leq C_2 k^2 h^l & \text{Dirichlet,} \\
E_h(k) &\leq C_2 k^3 h^{l+1} & \text{Neumann}
\end{aligned}
$$

where l is the degree of polynomial approximation for the boundary, and C_2 is a bounded constant, function of the boundary only. For the classical method, we can see that $k^{-1}E_h(k) \leq C_2 kh^l$ and $k^{-1}E_h(k) \leq C_2 k^2 h^{l+1}$ for the Dirichlet and Neumann boundary condition respectively. By the classical P^1 approximation $l = 1$ this error is easily controlled with the mesh criterion $hk \leq cst$, but this error remains unbounded with $hk^{1/3}$ even $l = 2$. As a high-order boundary approximation is not easy to perform and complicate the numerical computing, we've (naturally) abandon the idea of staying in one mesh level. Exactly we have:

- A mesh in size $h_g \sim k^{-1/3}$, on which we impose the degrees of freedom;

- A refined mesh in size $h_f \sim k^{-1}$, which approximate the real boundary. The finite element is first projected in this mesh and then the integrals evaluated.

With this 2 level mesh technique, we have the following estimates for the geometry approximations (see [15]):

$$
E_h(k) \leq C_2' k h_g^{-1} h_f^{l+1} \qquad\qquad \text{Dirichlet,}
$$

$$E_h(k) \leq C'_2 (kh_f + 1)(k^{1/2} h_g^{-1} h_f^l + k^{3/2} h_f^l) \qquad \text{Neumann}$$

We can take $l = 1$ for the Dirichlet case. Theoretically we must take $l \geq 2$ to the Dirichlet case (then Robin boundary condition case is the same error estimation as the Neumann case), but numerically $l = 1$ give also satisfactory results.

The case of a spherical incident wave is similar, we can also decompose the field in an analogous form, and it is for this reason that we write φ_0 instead of $(\vec{\nu} \cdot \vec{x})$.

6 Matrix calculation and resolution

We consider the continuous case, we must calculate integrals like

(6) $$\int_{\Gamma \times \Gamma} G(x, y) \, e^{ik(\varphi_0(y) - \varphi_0(x))} Z_r(x) \, Z_r(y) \, \tilde{q}(y) \, \lambda(x) \, d\gamma(x) d\gamma(y) \,,$$

(7) $$\int_{\Gamma \times \Gamma} \frac{\partial G}{\partial n_x}(x, y) \, e^{ik(\varphi_0(y) - \varphi_0(x))} Z_r(y) \, \tilde{q}(y) \, \lambda(x) \, d\gamma(x) d\gamma(y) \,,$$

and so on. As $G(r)$ is singular at $r = 0$, we split the integral into two parts by choosing a cut-off function ϕ, which is C^∞ and $\phi(x) = 1$ for $x \geq 2C$, $\phi(x) = 0$ for $x \leq C$ with C large enough. The regular integral (with a factor $\phi(k|x - y|)$ more) is performed first.

We consider here the case of dimension two, when $r > C$, the following asymptotic expansion of $H_\nu^{(1)}$ are used in (6) and (7) to simplify the calculation

$$H_\nu^{(1)}(r) \sim \sqrt{\frac{1}{2\pi r}} \, e^{i(r - \nu/2 - \pi/4)}, \qquad \nu = 0, 1$$

we obtain both for (6) and (7).

(8) $$\int_\Gamma \int_\Gamma M(x, y, k) \, e^{ik \, P(x,y)} \, d\gamma(y) d\gamma(x) \,,$$

with $\delta = 0$ for (6) and $\delta = 1$ for (7),

$$P(x, y) = |x - y| + (\varphi_0(y) - \varphi_0(x)) \qquad \text{and} \qquad M \in S_{2/3, 1/3}^{1/2 + \delta}((\Gamma \times \Gamma) \times \mathbb{R}) \,,$$

so the stationary phase method can be used.

The phase is stationary at points y where $\partial P / \partial \tau_y = 0$, that is

(9) $$\frac{(y - x, \tau_y)}{|x - y|} + \vec{\nabla}\varphi_0 \cdot \vec{\tau}(y) = 0 \,.$$

As $k\varphi_0$ is the first degree approximation of the phase function of q, it's just the phase of the incident wave, so φ_0 verifies the eikonal equation and $\vec{\nabla}\varphi_0$ is exactly the wave direction (here on Γ). The condition (9) admits a straightforward explanation : the point y is the reflection point (by Snell's law) of the transmission wave (*resp.* incident

8

wave) from which the reflected wave (*resp.* transmission wave) sticks x. The other condition $\partial P/\partial \tau_x = 0$ is analogous.

A little more investigation on P shows that (9) is only a necessary condition, there are points where $|\nabla_y^2 P| = 0$, so a direct application of the stationary phase theorem is not possible at these points.

We use the stationary phase formulas only for the inner integral of (8), for the reason that it is not sure that the phase is stationary on a finite number of $(x, y) \in (\Gamma \times \Gamma)$, (we don't want an object-dependent algorithm). For x fixed, on points y where $|\nabla_y^2 P| \neq 0$, the stationary phase formulas are used, whereas if $|\nabla_y^2 P| = 0$, we use the technics described in [7], Airy function and other special functions are involved. In practice, the inner integral is evaluated in each element of Γ_h, though for elements of Γ_h, $h \sim k^{-1/3}$. We can consider the element as infinite (there are $O(k^{2/3})$ periods for the phase function). But a direct use of the infinite domain stationary phase formulas in the program doesn't give a satisfactory result, and so the stationary phase formulas in bounded domains are used (cf. [7]). For the second integral (with respect to x), we must subdivide the segments to subelements of size $O(k^{-1})$, and integrate numerically on each. The total operation number for this part of integrals is

$$\underbrace{O(k)}_{\substack{\text{number of sub-}\\\text{domains for } x}} \quad \times \quad \underbrace{O(k^{1/3})}_{\substack{\text{number of}\\\text{elements of } \Gamma_h}} \quad = O(k^{4/3}) \,,$$

At dimension three, we can do the same thing for this part of integrals though the formulas will be more complex, the operation number will be $O(k^2)$.

We will see that this part of assembly time is small compared to that of the singular part, and recently a new way is in test to accelerate this calculation (assembly time expected to $O(k^{2/3})$ for 2D and $O(k^{4/3})$ for 3D for all the regular integrals).

The singular integral

$$\int_\Gamma \int_\Gamma Z_r(x) \, Z_r(y) \, G(k|x - y|) \, (1 - \phi(k|x - y|)) \, \tilde{q}(y) \, \lambda(x) \, d\gamma(x) d\gamma(y)$$

is computed in a costly manner, for the discrete formulas, the integral domains are divided into subdomains of size $h \sim k^{-1}$, and numerical integration is performed. For K, L two elements of Γ_h, if $\text{dist}(K, L) \geq c\,k^{-1/3}$, and $x \in K, y \in L$, $k|x - y| > ck^{2/3}$, so to each element K, only a few $(O(1))$ neighbor elements L need such a treatment, we have not yet found a rapid way to do this part of integrals. The assembly time for the singular integrals is, though the third factor can be replaced by $O(1)$:

$$\underbrace{O(k^{1/3})}_{\substack{\text{number of de-}\\\text{grees of freedom}}} \quad \times \quad \underbrace{O(k^{2/3})}_{\substack{\text{number of}\\\text{subdomains}}} \quad \times \quad \underbrace{O(k^{2/3})}_{\substack{\text{number of neighbor}\\\text{subdomains}}} \quad = O(k^{5/3}) \,,$$

in 2D and $O(k^{10/3})$ in 3D, however in the later case, due to the computer limitation, a very high frequency case can't be tested numerically.

9

The resolution time is not important, even with an ordinary Gauss algorithm, the time is of order $N^3 = O(k)$ in 2D and $N^3 = O(k^2)$ in 3D. Note that for the classical method, the assembly and resolution time are respectively $N^2 = O(k^2)$ and $N^3 = O(k^3)$ in 2D and $N^2 = O(k^4)$ and $N^3 = O(k^6)$ in 3D, the resolution time is more important in high frequency cases.

7 Numerical experiments

We present here some numerical results using this new method. Results are given to test the accuracy of this method. By physical optics, the fields in the "penumber" and dark region (see figure 1) depend essentielly on the current in the transition region, we decide to test our result in that region. As circles and spheres are only objects we can explicitly calculate the field developments for high frequency, we present at first the results in these cases.

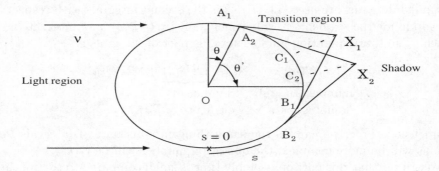

FIG. 1. *fields exterior to a circle*

We have the following asymptotic development for current in the transition region (*i.e* θ small)

$$(10) \qquad q(s) = \sum_{n=0}^{\infty} M_n(m, s, Z) \exp\left\{ iks + i\, k^{1/3}\, 2^{-1/3} \int_0^s \frac{\bar{\xi}_n(s)\, ds}{\rho(s)^{2/3}} \right\}.$$

The fisrt terms are respectively:

$$q_1(s) = 2^{1/3} k^{2/3}\, A(0)\, \mathrm{Ai}'(\xi_1) \exp\left\{ i\, k\, s + i\, k^{1/3}\, 2^{-1/3}\, \xi_1^d\, s \right\} \quad \text{Dirichlet,}$$

$$q_1(s) = A(0)\, \mathrm{Ai}(\xi_1^n) \exp\left\{ i\, k\, s + i\, k^{1/3}\, 2^{-1/3}\, \xi_1^n\, s \right\} \qquad \text{Neumann.}$$

Where Ai is the Airy function and ξ_1^d, ξ_1^n are the fisrt zero of Ai and Ai' respectively. We can express the modulus of the current in function of the angle θ, which gives,

$$\log\left(\left| \frac{q_1(s)}{q_1(0)} \right| \right) = -\alpha_1^{d,n}\, k^{1/3}\, \theta,$$

$$\alpha_d \sim 0.028 \qquad\qquad \alpha_n \sim 0.0122.$$

10

For θ' varying (in degree) from 75 to 80, we give the calculated constants α_d, α_n for different wavenumbers from $k = 400$ to $k = 1000$, all by respecting the mesh criterion $h = ck^{-1/3}$ (same c for all these wavenumber), in the following figures.

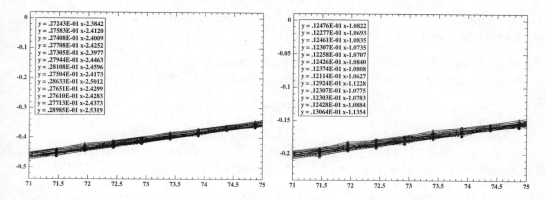

As indicated in [3], this method can apply also to irregular obstacles.

We refer to [15] for more results in dimension two and dimension three, regular and irregular obstacle cases, and also the importance of reffining the second mesh to obtain good results.

8   Conclusion

A new choice of the finite element space in using the boundary integral method has reduced the final unknown number to cube root of that as before, theoretically our method works for C^∞ convex objects, but in practice, that works also for piecewise C^2 smooth convex objects.

A second level of mesh is necessary to deduce the errors from the boundary approximation. This is essential to this method. Stationary phases method are used to evaluate the far field interaction, and the singular integral calculation becomes the most costly. This reduces the assembling time by a factor proportional to cube root of the wavenumber. The gain in assembling time is more important in high frequency than with a classical method. The total resolution time is proportional to the wavenumber, so becomes less important in the new method.

This singular integral calculation deserves amelioration.

References

[1] T. ABBOUD, J.C. NÉDÉLEC, B. ZHOU, *Improvement of the integral equation method for high frequency problems*, Third international conference on mathematical aspects of wave propagation phenomena, SIAM, p. 178-187, 1995.

[2] T. ABBOUD, J. EL GHARIB, J.C. NÉDÉLEC, T. SAYAH, An accelerated retarded potential method for time dependent acoustic wave scattering, *ASME 15th Biennal Conference on Mechanical Vibration and Noise*, Boston, Sept. 1995.

[3] K.R. ABEREGG, A.F. PETERSON, Application of the integral equation-asymptotic phase method to two-dimensional scattering, *IEEE Trans. on Ant. and Prop*, Vol. 43, No.5, p. 534-537, 1995.

[4] M. ABRAMOVITZ, I.A. STEGUN, *Handbook of mathematical functions*, National Bureau of Standards, Washington, 1970.

[5] F.X. CANNING, *The impedance matrix localization (IML) method for moment-method calculations IEEE Ant. and Prop.*, 18-30, Oct. 1990.

[6] D. COLTON, R. KRESS, *Integral equation methods in scattering theory*, Wiley & Sons, 1983.

[7] L. B. FELSEN, N. MARCUVITZ, *Radiation and Scattering of Waves*, chapter 4 : Asymptotic Evaluation of Integrals, Prentice-Hall, Englewood Cliffs, NJ 1973.

[8] J. GIROIRE, *Integral equation methods for exterior problems for the Helmholtz equation, in "Topics in differential and integral equations and operator theory"*, Vol. 5, Birkhäuser Verlag, Basel, 1982.

[9] M. HAMDI, *Une formulation variationnelle par équations intégrales pour la résolution de l'équation de Helmholtz avec conditions aux limites mixtes*, C. R. Acad. Sci. Paris, 292, Série II, 1981, p. 17-20.

[10] A. DE LA BOURDONNAYE, High frequency approximation of integral equations modeling scattering phenomena, *Math. Mod. and Num. Anal.*, Vol. 28, No. 2, p. 223-241, 1994.

[11] R.B. MELROSE, M.E. TAYLOR, Near peak scattering and the corrected Kirchhoff approximation for a convex obstacle, *Adv. in. Math.*, 55, 1985, p. 242-315.

[12] M.E. TAYLOR, *Pseudodifferential Operators*, Princeton University Press, 1981.

[13] F. TRÈVES, *Introduction to Pseudodifferential and Fourier Integral Operators*, Plenum Press, 1980.

[14] D. YINGST, *The kirchhoff approximation for Maxwell's equation*, Indiana Univ. Math. J. 32, 1983, No. 4, 543-562.

[15] B. ZHOU, *Méthode des équations intégrales pour la résolution des problèmes de diffraction à haute fréquence*, Thèse d'Université de Paris-XI, 1995.

GREGOIRE ALLAIRE and ALAIN ZELMANSE

Kinetic schemes for gas dynamics of real gases or two-phase mixtures

Abstract: Kinetic, or Boltzmann, schemes are numerical schemes based on a kinetic formulation of a hyperbolic system of conservation laws. For Euler equations describing the motion of a compressible inviscid fluid, there has been a renewed interest in this class of numerical algorithms after the recent work of Perthame. One drawback of his theory is that it applies only to so-called perfect gas which have a very simple thermodynamic state law. The goal of this paper is to present some numerical generalizations for real gas and equilibrated two-phase flows.

1 Introduction

This paper is devoted to the numerical study of so-called kinetic schemes for gas dynamics simulations of real gases and two-phase flows. Kinetic schemes (also called Boltzmann schemes) are a class of flux splitting schemes for the numerical solution of non-linear hyperbolic systems of conservation laws (see e.g. [9]). The main idea in a kinetic scheme is first to replace the original hyperbolic system by an equivalent transport equation for a distribution function depending on an additional variable (equivalent to a microscopic velocity), then to compute a solution of this transport equation, and finally to take moments of the distribution function to obtain approximate solutions of the original system. In the context of fluid mechanics this idea is very similar to the well-known formal asymptotics that leads to Euler equations when the mean free-path goes to zero in Boltzmann equation (see e.g. [1]). Recent works have shown that there exists an exact equivalence between some hyperbolic systems (mainly scalar conservation laws or the Euler isentropic system) and corresponding kinetic formulations which allows to prove new existence and stability results [12], [13], [14]. However, here we are only concerned with the application of such ideas to numerical computations. Although kinetic schemes have been used since the seventies, there has been recently a renewed interest in these schemes due to the fundamental work of Perthame [15], [16]. We shall follow his approach in the sequel.

Let us describe the kinetic approach to the numerical solution of hyperbolic systems of conservation laws. To simplify the presentation, we shall restrict ourselves to the one-dimensional case, but all the details go through in higher dimensions. Let us denote the time variable by $t \in \mathbb{R}^+$ and the space variable by $x \in \mathbb{R}$. Let F be a smooth function from $\mathbb{R}^p$ into $\mathbb{R}^p$, with $p \geq 1$. The unknown is a vector $u(t, x) \in \mathbb{R}^p$, solution of

$$\begin{cases} \frac{\partial u}{\partial t} + \frac{\partial F(u)}{\partial x} = 0 \\ u(t = 0, x) = u_0(x). \end{cases} \tag{1}$$

As is well-known, system (1) is supplemented with an entropy condition in order to

13

select a possibly unique relevant physical solution. In other words, for some convex function $S(u)$ (called the entropy) and some entropy flux $G(u)$ we ask the solution $u(t, x)$ to satisfy a further inequality $\frac{\partial S(u)}{\partial t} + \frac{\partial G(u)}{\partial x} \leq 0$.

The main idea is to introduce a new variable v (that can be thought of as a microscopic velocity) and a distribution function $f(t, x, v)$ (that can be interpreted as a particles density at time t, position x, and velocity v) such that its moments with respect to v are precisely equal to the solution $u(t, x)$ and to the flux $F(u(t, x))$

$$\int f(t, x, v) dv = u(t, x), \quad \int v f(t, x, v) dv = F(u(t, x)).$$

Remark that if u is a vector f may also need to be a vector, except if some other microscopic variables are introduced and each component of u is recovered by different moments of f (for example, an internal energy, see e.g. [15]). A kinetic formulation of the hyperbolic system (1) (if it exists) is a Boltzmann-type equation for the distribution function f

$$\frac{\partial f}{\partial t} + v \frac{\partial f}{\partial x} = Q(f), \tag{2}$$

where $Q(f)$ is a collision-type operator which forces f to relax to an equilibrium function $\mathcal{M}(x, v)$ which is the unique solution of the equation $Q(\mathcal{M}) = 0$. For the true Boltzmann equation, $\mathcal{M}(x, v)$ is called a Maxwellian and is a gaussian function in the v variable. The basis of kinetic schemes is **to solve in f rather than in** u and to recover the desired unknown u by taking moments of f.

As already said, a careful choice of Q allows in some cases to prove a strict equivalence between (1) and (2) (the entropy condition of (1) being recovered from the H-theorem for the kinetic equation (2)). However, for numerical purposes a precise form of the collision operator Q is not needed, and we simply require to know the equilibrium function $\mathcal{M}(x, v)$. Then, we can solve (2) by splitting the transport phase and the relaxation or collision phase. At each time step, we first solve the free transport equation

$$\frac{\partial f}{\partial t} + v \frac{\partial f}{\partial x} = 0,$$

then we apply the collision operator

$$\frac{\partial f}{\partial t} = Q(f).$$

Therefore, denoting by Δt the time step, a Boltzmann or kinetic numerical scheme can be decomposed in two steps

1. Convection (or free transport of f) with initial data $f^0(x, v)$

$$\begin{cases} \frac{\partial f}{\partial t} + v \frac{\partial f}{\partial x} = 0 & \text{for } 0 < t < \Delta t, \\ f(t = 0, x, v) = f^0(x, v) & \text{for } t = 0, \end{cases}$$

which admits the explicit solution $f(t, x, v) = f^0(x - tv, v)$.

2. Collision (or relaxation to equilibrium) : project the previous solution f onto the equilibrium function $\mathcal{M}(x,v)$ in order to obtain a new initial data $f^1(x,v)$ for the next convection step

$$f^1(x,v) = \mathcal{M}(x,v) \text{ with } \begin{cases} \int \mathcal{M}(x,v)dv = \int f(\Delta t, x, v)dv \\ \int v\mathcal{M}(x,v)dv = \int vf(\Delta t, x, v)dv \end{cases}$$

To define the collision step we implicitely assume that the equilibrium function $\mathcal{M}(x,v)$ is uniquely determined by its zero and first order moments in v.

Actually it turns out that, upon spatial discretization, a kinetic scheme simply reduces to a more classical flux splitting scheme. In other words, the v variable can be eliminated and a kinetic scheme can be implemented solely in terms of x and t. More precisely, discretizing the computational domains in cells of size Δx and denoting by $u_i(t)$ the average of $u(t,x)$ in cell i, the above kinetic scheme yields

$$u_i(\Delta t) = u_i(0) - \frac{\Delta t}{\Delta x} \left[F\left(u_i(0), u_{i+1}(0)\right) - F\left(u_{i-1}(0), u_i(0)\right) \right],$$

where $F(u_L, u_R)$ is the numerical flux function depending on the choice of the equilibrium function $\mathcal{M}(x,v)$. Different choices of the equilibrium function lead to different numerical schemes. In particular, if $\mathcal{M}(x,v)$ has a compact support in v, the resulting numerical scheme has a finite speed of propagation.

Kinetic schemes enjoy a few useful properties by their very definition. Since f satisfies a linear transport equation, it remains always positive if the equilibrium function $\mathcal{M}$ is positive. This yields a maximum principle for u (for example, in gas dynamics the density and pressure remain positive for all time). The so-called H-theorem for the kinetic equation (2) has its counterpart as a discrete entropy principle for the numerical scheme if the equilibrium function $\mathcal{M}$ is chosen as the minimum of some entropy minimization problem (see [15]). Kinetic schemes are naturally multidimensional by simply averaging a truely multi-dimensional kinetic free transport equation (one-dimensional splitting is not necessary). They can easily be extended to second order in space by a MUSCL approach.

In the context of the numerical simulation of Euler equations, describing the motion of an inviscid compressible gas, these ideas have already been explored by many individuals, including [3], [4], [5], [6], [10], [11], [15], [16], [17], [18], [19], and [21]. For scalar conservation laws, kinetic schemes have also been used in [2].

One drawback of kinetic schemes for gas dynamics is that they are only defined for so-called perfect gas which have a very simple thermodynamic state law. The goal of this paper is to try to generalize the kinetic approach to more general state laws of real gas and even to apply it for a homogeneous model of two-phase flow. The research described here is part of the PhD thesis of the second author [24].

2   Isentropic Euler equations

In this section we briefly recall two well-known kinetic schemes for the isentropic system of Euler equations

$$\begin{cases} \frac{\partial \rho}{\partial t} + \frac{\partial \rho u}{\partial x} &= 0 \\ \frac{\partial \rho u}{\partial t} + \frac{\partial (\rho u^2 + p)}{\partial x} &= 0 \end{cases} \tag{3}$$

supplemented with the perfect gas state law $p = \rho^\gamma$ with $\gamma > 1$, and the entropy condition $\frac{\partial S}{\partial t} + \frac{\partial Su}{\partial x} \leq 0$ with $S = \rho u^2 + \frac{2p}{\gamma(\gamma-1)}$. Each kinetic scheme is defined by an equilibrium function $\mathcal{M}(x,v)$ which will be the initial data of the linear transport equation $\frac{\partial f}{\partial t} + v\frac{\partial f}{\partial x} = 0$ solved between two consecutive time steps.

2.1   Kaniel scheme

The idea of Kaniel [10] for computing numerical solutions of (3) is to use the following type of equilibrium function

$$\mathcal{M}(x,v) = \chi(|v - u|)1_{|v-u|<\alpha(\rho)},$$

where the shape χ of the equilibrium function and the cut-off function α depends only on the pressure law, i.e. on γ. In order to recover the macroscopic unknowns of (3), we ask that $\mathcal{M}$ satisfies

$$\int_{|v|<\alpha(\rho)} \chi(|v|)dv = \rho. \tag{4}$$

Then, automatically since $\chi(|v|)$ is even, one has

$$\int \mathcal{M}(x,v)dv = \rho(x) \text{ and } \int v\mathcal{M}(x,v)dv = \rho(x)u(x).$$

Condition (4) yields also the consistency with the first equation of (3) (conservation of mass) since, up to first order in time, it is obtained by averaging the linear transport equation with respect to v. In order to obtain the consistency with the second equation (conservation of momentum), we add another constraint

$$\int_{|v|<\alpha(\rho)} |v|^2\chi(|v|)dv = p. \tag{5}$$

Conditions (4) and (5) uniquely determine both functions χ and α. A simple computation yields

$$\alpha(\rho) = \sqrt{\frac{\partial p}{\partial \rho}} = \sqrt{\gamma\rho^{\gamma-1}}, \text{ and } \chi(\alpha) = \alpha\left(\frac{\partial^2 p}{\partial \rho^2}(\alpha)\right)^{-1} = \frac{\alpha^{\frac{3-\gamma}{\gamma-1}}}{(\gamma-1)\gamma^{\frac{1}{\gamma-1}}}.$$

This scheme is well-defined for any value of $\gamma > 1$, even when $\gamma > 3$. The width of the support of the Kaniel equilibrium function is exactly equal to the sound speed (at least in 1-D). However, we are unable to prove any discrete entropy property for this scheme (except when $\gamma = 3$).

16

2.2 Perthame scheme

The idea of Perthame [15] is to obtain an equilibirum function

$$\mathcal{M}(x, v) = \chi(|v - u|),$$

where $\chi(v)$ is a minimizer of a kinetic entropy defined by

$$H(f) = \int \left(\kappa f(v)^{\frac{\gamma+1}{3-\gamma}} + v^2 f(v) \right) dv$$

among all possible distribution functions f satisfying the constraints

$$f \geq 0, \quad \int f(v) dv = \rho, \quad \int v^2 f(v) dv = p.$$

The constant κ is chosen precisely equal to $\frac{2(3-\gamma)}{(\gamma-1)^2}(8I_\gamma/(\gamma-1))^{(\gamma-1)/(3-\gamma)}$ with $I_\gamma = \int_0^1 (1 - t^2)^{\frac{3-\gamma}{2(\gamma-1)}} dt$. For $1 < \gamma \leq 3$ this is a convex minimization problem which admits a unique minimizer given by

$$\chi(v) = \frac{1}{2I_\gamma} \sqrt{\frac{\gamma - 1}{2}} \left(\rho^{\gamma-1} - \frac{\gamma - 1}{2}(v - u)^2 \right)_+^{\frac{3-\gamma}{2(\gamma-1)}},$$

where $(\cdot)_+$ denotes the positive part function. Clearly, because of the constraints the Perthame equilibrium function is consistent with the system (3). Furthermore, due to the special choice of the constant κ, an easy computation shows that $H(\chi) = S$. Since the kinetic entropy $H(f)$ is simply transported as f during the free transport step and then is decreased during the collision step because $H(\chi) \leq H(f)$, Perthame scheme satisfies a corresponding discrete entropy principle.

Let us remark that when $\gamma = 3$ Perthame and Kaniel schemes coincide and the equilibrium function is a simple step function. When $\gamma > 3$, Perthame scheme is not any longer entropic (i.e. the minimum of the kinetic entropy is $-\infty$), but the formula for χ still yields a well-defined numerical scheme (that actually behaves poorly).

3 Euler equations

In this section we introduce a family of Perthame kinetic schemes for the full system of Euler equations

$$\begin{cases} \frac{\partial \rho}{\partial t} + \frac{\partial \rho u}{\partial x} &= 0 \\ \frac{\partial \rho u}{\partial t} + \frac{\partial (\rho u^2 + p)}{\partial x} &= 0 \\ \frac{\partial E}{\partial t} + \frac{\partial (E+p)u}{\partial x} &= 0 \end{cases} \tag{6}$$

with the total energy $E = 1/2\rho u^2 + \rho e$ and the perfect gas state law $p = (\gamma - 1)\rho e$ with $\gamma > 1$, and the entropy condition $\frac{\partial S}{\partial t} + \frac{\partial Su}{\partial x} \leq 0$ for $S = \rho log \left(\frac{e}{\rho^{(\gamma-1)}} \right)$.

To define a kinetic scheme for system (6) another microscopic variable is required (a particle internal energy) or another distribution function. The second choice is more convenient and we work with two equilibrium functions $\mathcal{M}(x,v)$ and $\mathcal{N}(x,v)$ which will be the initial data of two linear transport equations of the type $\frac{\partial f}{\partial t} + v\frac{\partial f}{\partial x} = 0$ at each time step. For $1 < \gamma \le 3$, the couple $(\mathcal{M}, \mathcal{N})$ is defined by

$$\mathcal{M}(x,v) = \chi(|v - u(x)|) \text{ and } \mathcal{N}(x,v) = \tilde{\chi}(|v - u(x)|),$$

where $(\chi, \tilde{\chi})$ is the unique minimizer of the convex kinetic entropy

$$H(f,g) = \int f(v)^p g(v)^q dv$$

among all possible non-negative distribution functions (f,g) satisfying the constraints

$$\int f(v)dv = \rho, \quad \int \left(\frac{v^2}{2} f(v) + g(v) \right) dv = e,$$

with $p > 1$ and $q = (p-1)\frac{\gamma-3}{\gamma+1}$. This choice of q ensures the consistency with system (6), i.e. that $\int v^2 f(v)dv = p$, and that $H(\chi, \tilde{\chi})$ is precisely an entropy of system (6). Defining $\lambda = \frac{q-1}{1-p-q}$ the minimizer is given by

$$\chi(x,v) = C_1 \frac{\rho}{e} \left(1 - C_2 \frac{v^2}{\rho^2} \right)^{\lambda}_+, \quad \tilde{\chi}(x,v) = C_3(e - C_4\rho^2)\chi(x,v)^{\frac{p}{1-q}},$$

where C_1, C_2, C_3, C_4 are positive constants (see [24] for their precise values). The case $p = +\infty$ is described in [11]. This choice of kinetic entropy $H(f,g)$ has the very interesting feature of giving rise to compactly supported equilibrium functions. This is a highly desirable property for numerical purposes since the resulting scheme will be less diffusive. In particular, a compact support in v implies a finite speed of propagation for the numerical scheme.

The kinetic scheme associated to the equilibrium functions $(\chi, \tilde{\chi})$ is well-defined and entropic for $1 < \gamma \le 3$. For $\gamma = 3$ it is very simple since $\tilde{\chi} = 0$ and χ is a step function. It is not defined for $\gamma > 3$.

4   Extension to real gas and two-phase mixtures

Most kinetic schemes are designed only for perfect gas, i.e. for a state law of the type $p = (\gamma - 1)\rho e$ with a constant $\gamma > 1$. Indeed, for the kinetic schemes described above, the constant γ is built in the equilibrium functions which in turn yields the numerical flux formula. Our goal here is to extend, at least from a numerical point of view, these kinetic shemes for real gas which may have very complicated state laws.

Part of our motivation is also to extend Boltzmann schemes for some two-phase flow models. Here we shall be concerned only with the so-called homogeneous equilibrated two-phase flow model which describes the motion of a mixture of liquid and vapor. For extensions to another two-phase flow model, the so-called bi-fluid model, we refer to [7]. The homogeneous equilibrated two-phase flow model is the system of Euler equations

18

for the mixture with a diphasic state law which is very stiff and has discontinuous derivatives when phase changes occur. The main physical assumptions allowing to derive this model are that the two phases must be thermodynamically equilibrated and have the same speed, the same pressure, and the same temperature. A typical state law for a two-phase mixture of water and steam is that of the benchmark [8] (see figure 1). In the liquid region the mixture is almost incompressible, in the diphasic region it is almost isothermal, and in the vapor region it behaves like a real gas. Obviously it is a very severe example of real gas !

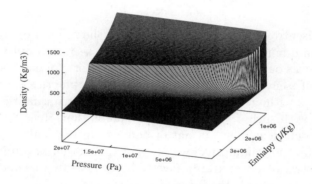

FIG. 1. State law $\rho(p, h)$

For such two-phase flow problems, many numerical schemes are available, but the state of the art is probably provided by Toumi's adaptation of Roe scheme [23]. The main trick is that the linearization depends on the thermodynamic states considered, and therefore contains more physics than a simple upwind scheme. Despite its advantages the modified Roe scheme is not the ultimate scheme, and better ones are always sought. In particular, kinetic schemes have the following potential advantages : positivity of important variables, like density or pressure, is preserved (in the context of two-phase flows this could be helpful for the concentration to stay between 0 and 1), discrete entropy properties are easily established, and finally kinetic schemes have a natural multidimensional character.

Our strategy for extending the classical kinetic schemes to real gas and two-phase mixtures is to use a so-called *effective γ-coefficient technique*. In each cell and at each time step, for given thermodynamic conditions we shall compute a constant γ such that the true state law is equivalent to a perfect gas state law ($p = (\gamma - 1)\rho e$). Then, locally in space and time the numerical fluxes are computed with one of the above kinetic scheme for this value of γ. In other words, the numerical fluxes depend on γ and are therefore a nonlinear function of the thermodynamic variables. We consider two possible equivalences of the true state law and that of a perfect gas. First, we simply compute γ by the obvious formula

$$\gamma = \frac{p}{\rho e} + 1, \tag{7}$$

and we use a kinetic scheme for the full system of Euler equations. An alternative is to assume that the flow is isentropic and to approximate the true state law by that of an isentropic perfect gas ($p = A\rho^\gamma$). In this case we compute two constants A and γ such that the pressure and the sound speed ($c = \sqrt{\frac{\partial p}{\partial \rho}(\rho, S)}$) coincide

$$\gamma = \frac{\rho c^2}{p}, \ \ A = \frac{p}{\rho^{\frac{\rho c^2}{p}}}.$$

Then, we use an isentropic kinetic scheme, but for the full system of Euler equations (i.e. two equilibrium functions are introduced like in section 3 but their shape is that of section 2). It turns out that this last procedure is more robust than the former one. The assumption of isentropicity is reasonable since in practice we consider only low Mach number flows without shocks (but with contact discontinuities).

The difficulty with such a γ effective technique is that, due to the stiffness of the state law, the γ coefficient may run out of bounds. In particular, the incompressible limit (when the fluid is a liquid like water) and the isothermal limit (when the fluid is an equilibrated two-phase mixture) have to be investigated to understand potential numerical degeneracy. This is the focus of next section.

5 Various asymptotic limits

This section is devoted to the study of asymptotic limits of the kinetic fluxes when γ is close to some of its imposed bounds. For example, recall that γ must always be larger than 1 and smaller than 3 (in 1-D) to ensure that a kinetic scheme is entropic.

5.1 Incompressible limit for Euler equations

The incompressible limit is equivalent to that of low Mach number, i.e. the sound speed is much larger than the average fluid velocity. For a perfect gas, this implies that γ is close to 1. A careful inspection of the formula of section 3 shows that the equilibrium distribution function χ and $\tilde{\chi}$ converge to Dirac masses, and that upon a suitable rescaling they locally behave like a gaussian near the origin. This shows that the numerical fluxes are almost degenerate for γ close to 1, and that a potential cure is to use a gaussian distribution function (which, as is well-known, gives rise to a very diffusive scheme).

20

In the case of the isentropic system of Euler equations, the incompressible limit (which is still equivalent to a very large sound speed corresponds to very large values of γ (much bigger than 3). Clearly, the Perthame scheme is not very good in this regime : it is not isentropic and the equilibrium function blows up at both ends of its support. The Kaniel scheme seems to be not much better, its equilibrium function blows up too, but numerically it behaves in a better way for large γ (for details, see [24]).

5.2   Isothermal limit for isentropic Euler equations

For a two-phase mixture as described in the previous section, there is another asymptotic regime in the diphasic region of the state law. Indeed, assuming thermodynamic equilibrium, if the two phases cohabit, their thermodynamical properties must lie on the saturation curve. In other words, for each individual phase the knowledge of the pressure is enough to determine all other quantities (temperature, enthalpy, etc.). The other variable is then the concentration of one phase. In practice it is almost an isothermal process. For an isentropic perfect gas, this implies that γ is close to 1 in the diphasic regime. When γ tends to 1, the equilibrium function of Perthame scheme converges to a gaussian

$$\lim_{\gamma \to 1} \chi(v) = \frac{1}{2\sqrt{2\pi}} \rho e^{-\frac{1}{2}v^2}.$$

Of course, this implies that, in the limit, the scheme is very diffusive (we refer to [24] for attempts to use a truncated gaussian in order to obtain a better kinetic scheme in this limit).

6   Shock tube in the incompressible limit

We have performed a shock tube test problem for the state law of [8] (see figure 1). The full system of Euler equations is used, and the fluid remains liquid (almost incompressible) through the all test. The precise intitial conditions may be found in [24], and we simply remark that it features a strong shock for a liquid (more than two bars). We compare the results of Perthame scheme with first a fixed value of $\gamma = 3$ (step function), and second a varying γ fitted through (7) to a perfect gas state law. In the latter case, the effective γ runs between 1.008 and 1.006. Remark that the time step must be smaller than the CFL condition. Of course, the best results stem from the varying effective γ strategy. Good results are also obtained with Kaniel scheme for which the numerical speed of propagation coincides with the sound speed.

7   Two-phase flow test

We also performed a classical benchmark in two-phase flow : the first CSNI Numerical Benchmark Problem [8]. It is a vertical tube filled initially with water that is heated on part of its length. The water begins to boil, preventing more water to enter the tube, but after the vapor has exited the tube, new water gets in and begin to boil.

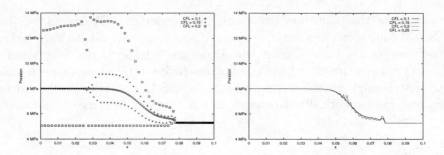

FIG. 2. Pressure, step function (left), entropic scheme (right)

This cycle repeats in time and is damped only by dissipation and friction against the tube walls. The purpose of the test is to compute oscillations that are not damped too quickly by the numerical dissipation. We compare our solution to that of Toumi using a modified Roe scheme [23]. Remark that we need to use a finer grid with the kinetic sceme (Perthame scheme with effective γ). Nevertheless the results are worse : the oscillations are damped too much and their frequency is not correct.

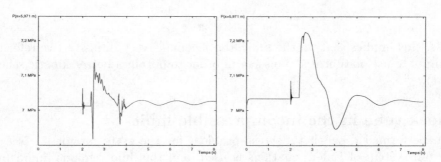

FIG. 3. Outlet pressure as a function of time. Modified Roe scheme of Toumi, 60 cells (left), kinetic scheme (source terms implicited), 300 cells (right)

As a conclusion we can say that kinetic schemes behave poorly for equilibrated two-phase flow (but Lax-Friedrichs scheme is even worse). Remark that the state law is very stiff and features discontinuities of thermodynamic derivatives at phase changes. However, the main problem with kinetic schemes is that they are known to be very diffusive through contact discontinuities which are the difficulties in the present subsonic situation. Notice also that our adaptation of kinetic schemes to general state laws is completely blind to any physical available information. This is in sharp contrast with the two-phase flow version of Roe scheme devised by Toumi [23] which uses the saturation curves to compute the linearized jacobian in the diphasic regime. Let us mention that kinetic schemes may be more adapted to two-fluid models (not equilibrated) [7].

22

References

[1] BARDOS C., GOLSE F., LEVERMORE D., *Macroscopic limits of kinetic equations,* Multi-dimensional hyperbolic problems and computations, pp.1-12, North-Holland (1993).

[2] BRENIER Y., *Averaged multivalued solutions for scalar conservation laws,* SIAM J. Num. Anal. 21, pp.1013-1037 (1984).

[3] CORON F., PERTHAME B., *Numerical passage from kinetic to fluid equations,* SIAM J. Num. Anal. 28, pp.26-42 (1991).

[4] CROISILLE J.-P., DELORME P., *Kinetic symmetrizations and pressure laws for the Euler equations,* Physica D, 57, pp.395-416 (1992).

[5] DESHPANDE S.M., *A second-rder accurate kinetic-theory-based method for inviscid compressible flows,* NASA Technical Papers 2613 (1986).

[6] DE VUYST F., *Un nouveau schéma pour la résolution des équations d'Euler compressibles,* C. R. Acad. Sci. Paris 314, pp.577-582 (1992).

[7] EL AMINE K., *Modélisation et analyse numérique des écoulements diphasiques en déséquilibre,* Thèse de l'Université Paris 6 (1997).

[8] Energie Atomique du Canada, *First CSNI Numerical Benchmark Problem,* Etablissement de Recherches Nucléaires de Whiteshell, TRB-79-30 (1979).

[9] HARTEN A., LAX P., VAN LEER B., *On upstream differencing and Godunov-type schemes for hyperbolic conservation laws,* SIAM Review 25, pp.35-61 (1983).

[10] KANIEL S., *A kinetic model for the compressible flow equations,* Indiana Univ. Math. Journal 37, pp.537-563 (1988).

[11] KHOBALATTE B., PERTHAME B., *Maximum principle on the entropy and minimal limitations for kinetic schemes,* Mathematics of Computations 62, pp.119-131 (1994).

[12] LIONS P.-L., PERTHAME B., TADMOR E., *A kinetic formulation of multi-dimensional scalar conservation laws and related equations,*

[13] LIONS P.-L., PERTHAME B., TADMOR E., *Kinetic formulation of the isentropic gas dynamics and p-systems,* Comm. Math. Physics 163, pp.415-431 (1994).

[14] LIONS P.-L., PERTHAME B., SOUGANIDIS P., *Existence and stability of entropy solutions for the hyperbolic systems of isentropic gas dynamics in Eulerian and Lagrangian coordinates,* Publication du Laboratoire d'Analyse Numérique de l'Université Paris 6, R.95004 (1995).

[15] PERTHAME B., *Boltzmann type schemes for gas dynamics and the entropy property*, SIAM J. Num. Anal. 27, pp.1405-1421 (1990).

[16] PERTHAME B., *Second-order Boltzmann schemes for compressible Euler equations in one and two space dimensions*, SIAM J. Num. Anal. 29, pp.1-19 (1992).

[17] PRENDERGAST K., XU K., *Numerical hydrodynamics from gas kinetic theory*, J. Comp. Phys. 109, pp.53-66 (1993).

[18] PULLIN D., *Direct simulation methods for compressible inviscid ideal-gas flow*, J. Comp. Phys. 34, pp.231-244 (1980).

[19] REITZ R., *One-dimensional compressible gas dynamics calculations using the Boltzmann equation*, J. Comp. Phys. 42, pp.108-123 (1981).

[20] ROE. P.-L., *Approximate Riemann solvers, parameter vectors, and difference schemes*, J. Comp. Phys. 43, pp.357-372 (1981).

[21] SANDERS R., PRENDERGAST K., *The possible relation of the 3-kiloparsec arm to explosions in the galaxic nucleus*, The Astrophysical Journal 188, pp.489-500 (1974).

[22] STEWART H., WENDROFF B., *Two-phase flow: models and methods*, J. Comp. Phys. 56, pp.363-409 (1984).

[23] TOUMI I., *A weak formulation of Roe's approximate Riemann solver*, J. Comp. Phys. 102, pp.360-373 (1992).

[24] ZELMANSE A., *Formulation cinétique et schémas de Boltzmann pour le calcul numérique en mécanique des fluides*, Thèse de l'Université Paris 13 (1995).

LILLIAM ALVAREZ, ANTONIO RODRIGUEZ and IGNACIO SANCHEZ

A numerical technique to solve linear and non-linear singularly perturbed problems

1 Introduction

Many problems in Fluid Mechanics are sources of mathematical models in which coefficients appear big parameters, for example, the Reynolds number $(R_e >> 1)$ in hidrodinamical problems.

In this paper we present an efficient numerical technique to solve scalar singular perturbation problems for second order ordinary differential equations (ODEs) of type:

$$\varepsilon y' = f(x, y, y'), \qquad a < x < b$$

$$g(a, y(a), y'(a)) = \gamma_1$$

$$h(b, y(b), y'(b)) = \gamma_2$$

where f, g, h are non-linear functions and $\varepsilon << 1$.

In the case of PDEs, the method of lines is applied to reduce the model to a system of ODEs of the previous type.

The numerical technique described in this work was firstly proposed by Abramov and Alvarez in 1989 [1], and will be presented in section 2. The main idea is to splitt the differential operator in a sequence of stiff initial value problems, wich will be solved using a semi-analytical quadrature formulae, also given in [1] and formulated in section 3.

The application of these both techniques: operator splitting and quadrature formulae to solve test problems like some non-linear singularly perturbed problem for ODEs, the Orr-Sommerfield equation, and the one-dimensional Stefan problem for one and two phases are shown in section 4.

2 The operator splitting method (OSM)

For a general linear two-point boundary value problem

$$\varepsilon y''(x) + p(x, \varepsilon) y'(x) + q(x, \varepsilon) y(x) = f(x, \varepsilon) \qquad (2.1.a)$$

$$\alpha_1 y(a) + \beta_1 y'(a) \qquad = \gamma_1(\varepsilon) \qquad (2.1.b)$$

$$\alpha_2 y(b) + \beta_2 y'(b) = \gamma_2(\varepsilon) \qquad (2.1.c)$$

25

where $0 < \varepsilon << 1$, if D and D^2 denotes the first and second derivates, then $(2.1.a)$ can be written as

$$\left(D^2 + \frac{p(x,\varepsilon)}{\varepsilon}D + \frac{q(x,\varepsilon)}{\varepsilon}\right) y = \frac{f(x,\varepsilon)}{\varepsilon} \qquad (2.2)$$

and its characteristic equation is

$$\mu^2 + \frac{p(x,\varepsilon)}{\varepsilon}\mu + \frac{q(x,\varepsilon)}{\varepsilon} = 0 \qquad (2.3)$$

whose roots are

$$\mu_{1,2}(x,\varepsilon) = \frac{-\frac{p(x,\varepsilon)}{\sqrt{\varepsilon}} \pm \sqrt{\frac{p^2(x,\varepsilon)}{\varepsilon} - 4q(x,\varepsilon)}}{2\sqrt{\varepsilon}} \qquad (2.4.a)$$

and we set

$$\mu_{i0}(x,\varepsilon) = \sqrt{\varepsilon}\mu_i(x,\varepsilon) \qquad \text{for } i = 1,2 \qquad (2.4.b)$$

Then by means of the characteristic roots, if $\mu_1$exists in $[a,b]$ we have

$$(D - \mu_2(x,\varepsilon))(D - \mu_1(x,\varepsilon)) y = \frac{f(x,\varepsilon)}{\varepsilon} - \mu_1(x,\varepsilon) y \qquad (2.5)$$

multiplying (5) by ε, considering $(4.b)$ and denoting

$$z = \sqrt{\varepsilon}(D - \mu_1(x,\varepsilon)) y \qquad (2.6)$$

we obtain the equivalent system

$$\sqrt{\varepsilon}z' - \mu_{20}(x,\varepsilon) z = f(x,\varepsilon) - \sqrt{\varepsilon}\mu_{10}(x,\varepsilon) y \,(2.7.a)$$

$$\sqrt{\varepsilon}y' - \mu_{10}(x,\varepsilon) y = z(x) \qquad (2.7.b)$$

It is easy to see if z_0, z_1, y_0, $y_1$and y_2 are complex functions such that satisfy the following problems

$$\sqrt{\varepsilon}z'_0 - \mu_{20}(x,\varepsilon) z_0 = f(x,\varepsilon) - \sqrt{\varepsilon}\mu_{10}(x,\varepsilon) y, \; z_0(a) \text{ or } z_0(b) = 0$$
$$(2.8.a)$$

$$\sqrt{\varepsilon}z'_1 - \mu_{20}(x,\varepsilon) z_1 = 0, \; z_0(a) \text{ or } z_o(b) = 1 \qquad (2.8.b)$$

$$\sqrt{\varepsilon}y'_0 - \mu_{10}(x,\varepsilon) y_0 = z_0, \; y_0(a) \text{ or } y_0(b) = 0 \qquad (2.8.c)$$

$$\sqrt{\varepsilon}y_1' - \mu_{10}(x, \varepsilon) y_1 = z_1, \quad y_1(a) \text{ or } y_1(b) = 0 \tag{2.8.d}$$

$$\sqrt{\varepsilon}y_2' - \mu_{10}(x, \varepsilon) y_2 = 0, \quad y_2(a) \text{ or } y_2(b) = 1 \tag{2.8.e}$$

the solution of (1) can be rewritten as

$$y(x) = \text{Re}\left(y_0(x) + C_1 y_1(x) + C_2 y_2(x)\right) \tag{2.9}$$

where C_1 and C_2 can be found from $(2.1.b)$ and $(2.1.c)$, solving a linear system. Here Re represents the real part of complex number.

We choose the condition in a or b according to the sign of the real part of $(2.4.a)$ in order to get stable initial value problems.

Observing $(2.8.a)$, we note that the right hand side of the equation depends on the solution of the original problem (1). We solve this difficulty by means of an iterative process as follows:

Step 1. We solve the problem $(2.8.b)$, $(2.8.d)$ and $(2.8.e)$.

Step 2. Given an approximation for y, we solve the problem $(2.8.a)$.

Step 3. Given z_0, we solve the problem $(2.8.c)$.

Step 4. We compute C_1 and C_2 solving the linear system which results from $(2.1.b)$, $(2.1.c)$ and (2.9)

Step 5. We form a new approximation of y by (2.9) and if the imposed converge condition is not fulfilled, we go back to step number two.

It is easy to see, that (OSM) can be applied if the boundary conditions are non-linear. In this case we only have to solve a non-linear algebraic system instead of linear one for the constants C_1 and C_2.

The convergence analysis of this iterative variant was reported [2], and it gave that, under certain conditions the stiffer of (2.1) the faster algorithm converges.

3 Quadrature formulae for the Initial Value Problems (IVPs)

We are left with the numerical solution of the problems (2.8). In this section we will obtain explicit quadrature formulas [7], that allows us to solve efficiently our auxiliary Stiff Cauchy problems.

For the problem

$$\sqrt{\varepsilon}u' + \lambda(x) u = v(x), \quad u(0) = 0 \tag{3.1}$$

noindentwhere $\text{Re}\,\lambda(x) \geq \alpha_0$ and $\varepsilon << 1$.

Define a grid $0 = x_0 < ... < x_N = a$ in $[0, a]$. In what follows, for any function $f(x)$, we put $f_k = f(x_k)$. We need to compute $u_1, ..., u_N$, while $v_0, ..., v_N, \lambda_0, ..., \lambda_N$ and $\Lambda_0, ..., \Lambda_N$ are known; here

$$\Lambda(x) = \int_0^x \lambda(z)\,dz$$

The computations are carried out from left to right. We will give a recursive formulae that expresses u_{k+1} in terms of u_k and the given quantities.

From (3.1) for $k = 0, 1, \ldots, N-1$ we have

$$u_{k+1} = u_k \exp\left[-\sqrt{\varepsilon}\left(\Lambda_{k+1} - \Lambda_k\right)\right] + \phi_k \qquad (3.2)$$

where

$$\phi_k = \int_{x_k}^{x_{k+1}} \exp\left\{-\sqrt{\varepsilon}\left[\Lambda_{k+1} - \Lambda(t)\right]\right\} \frac{v(t)}{\lambda(t)} d\left\{-\sqrt{\varepsilon}\left[\Lambda_{k+1} - \Lambda(t)\right]\right\}$$

Let us denote $\varphi(t) = \frac{v(t)}{\lambda(t)}$. Let $v(t)$ be a "good" function. Represent $\varphi(t)$ in the form

$$\varphi(t) = \varphi_{k+1} + \frac{\Lambda_{k+1} - \Lambda(t)}{\Lambda_{k+1} - \Lambda_k}\left[(\varphi_{k+1} - \varphi_k) + o\left[(t_{k+1} - t_k)^2\right]\right] \quad (3.3)$$

then for the principal part of $\varphi(t)$, the integration can be carried out explicitly, and we obtain

$$\phi_k \approx \frac{v_{k+1}}{\lambda_{k+1}}\left\{1 - \exp\left[-\sqrt{\varepsilon}\left(\Lambda_{k+1} - \Lambda_k\right)\right]\right\} + \frac{\frac{v_k}{\lambda_k} - \frac{v_{k+1}}{\lambda_{k+1}}}{\sqrt{\varepsilon}\left(\Lambda_{k+1} - \Lambda_k\right)}$$
$$\left\{1 - \left[1 + \sqrt{\varepsilon}\left(\Lambda_{k+1} - \Lambda_k\right)\right]\exp\left[-\sqrt{\varepsilon}\left(\Lambda_{k+1} - \Lambda_k\right)\right]\right\}$$

From (3.2) we now obtain the required recursive formulae

$$u_{k+1} = u_k \exp\left[-\sqrt{\varepsilon}\left(\Lambda_{k+1} - \Lambda_k\right)\right]$$
$$+ \frac{v_{k+1}}{\lambda_{k+1}}\left\{1 - \exp\left[-\sqrt{\varepsilon}\left(\Lambda_{k+1} - \Lambda_k\right)\right]\right\} + \frac{\frac{v_k}{\lambda_k} - \frac{v_{k+1}}{\lambda_{k+1}}}{\sqrt{\varepsilon}\left(\Lambda_{k+1} - \Lambda_k\right)}$$
$$\left\{1 - \left[1 + \sqrt{\varepsilon}\left(\Lambda_{k+1} - \Lambda_k\right)\right]\exp\left[-\sqrt{\varepsilon}\left(\Lambda_{k+1} - \Lambda_k\right)\right]\right\} \quad (3.4)$$

which has the following properties:
a) it is stable for any x-step and ε (provided of course, $\mathrm{Re}\,\lambda(x) \geq \alpha_0 > 0$);
b) it gives a global error

$$O\left(h\min\left(h, \sqrt{\varepsilon}\right)\right) \text{ on } [0, a], \text{ where } h = \max_k \left(x_{k+1} - x_k\right)$$

If $\mathrm{Re}\,\lambda(x) \geq \alpha_0 > 0$ an analogous backward formulae can be obtained.

28

This result was obtained by integrating the error in (3.3) over the interval $[x_{k+1}, x_k]$ and summing over all intervals.

For $\varphi(t)$ we used an interpolation formulae linear in $\Lambda(t)$. If (3.3) is replaced with an expression which is a polynomial of higher degree in $\Lambda(t)$, then we obtain an explicit formulae with property a) that has an accuracy of higher order (an analogue of Adams formulas).

4 Numerical experiments

All the following examples were solved in a PC, using a very simple FORTRAN 77 code. In all cases the exact numerical error was less than 10^{-4}. The execution time was always of a few seconds.

4.1 Test problems

4.1.1 Boundary Layer Problems:[1], [5], respectively:

$$\varepsilon y'' - xy' - \frac{1}{2}y = 0$$

$$y(-1) = 1 \qquad y(1) = 2$$

This example was solved for $\varepsilon = 10^{-3}$, $N = 121$, relative error$= 10^{-6}$ and absolute error$= 10^{-10}$.

$$\varepsilon y'' - (2 + \cos(\pi x)) y' - y = F(x)$$

$$y(0) = 0 \qquad y(1) = -1$$

where

$$F(x) = -\left(1 + \varepsilon n^2\right) \cos(\pi x) - \pi \left(2 + \cos(\pi x)\right) \sin(\pi x) + \left(1 + \frac{\pi^2 x^2}{\varepsilon}\right) e^{-\frac{3x}{e}}$$

Its exact solution is

$$y(x) = \cos(\pi x) - e^{-\frac{3x}{e}} + o\left(\varepsilon^2\right)$$

This example was solved for $\varepsilon = 10^{-6}$, $N = 121$, the maximal order with respect to the exact solution was 10^{-4}.

4.1.2 Turning Point Problems, [3]

$$\varepsilon y'' - xy' - \frac{1}{2}y = 0$$

$$y(-1) = 1 \qquad y(1) = 2$$

Its turning point is for $x = 0$. We took $\varepsilon = 10^{-3}$, $N = 221$, relative error= 10^{-6} and absolute error= 10^{-10}.

$$\varepsilon y" - xy' - y = -\left(1 + \varepsilon n^2\right) \cos(\pi x) - \pi x \sin(\pi x)$$

$$y(-1) = -1 \qquad y(1) = 1$$

Its turning point is for $x = 0$ also, and was solve for $\varepsilon = 10^{-3}$, $N = 241$, relative error= 10^{-6} and absolute error= 10^{-10}.

4.1.3 Semilinear Singularly Perturbed Problems

$$\varepsilon y" - xy' - y(1 - y) = 0$$

$$y(-1) = 0 \qquad y(1) = 1$$

In this case was taken $\varepsilon = 10^{-3}$, relative error= 10^{-6} and absolute error= 10^{-10}.

$$\varepsilon y" - xy' - y\left(y^2 - 1\right)(2 - y) = 0$$

$$y(-1) = -1 \qquad \text{and} \qquad y(1) = 1$$

We took $\varepsilon = 10^{-3}$, relative error= 10^{-6} and absolute error= 10^{-10}.

4.1.4 Rapidly Oscillatory Problems
To solve this kind of problem, a non-iterative variant of the (OSM) was proposed in [6] using the Riccati transformation. This is the variant used to solve also the next non-linear test.

$$\varepsilon y" - \left(\varepsilon a t^2 + 1\right) y = 0$$

$$y(0) = 1 \qquad \text{and} \qquad y(5) = 0$$

$$a = 0.25$$

4.1.5 Non Linear Problem

$$\varepsilon y" - \frac{1}{4} yy' - y = 0$$

$$y(-1) = -1 \qquad \text{and} \qquad y(1) = 2$$

4.1.6 Stefan Problem

This problem appears in many fields of the science when we study phenomena like evaporation, fusion, solidification, etc. An important characteristic of this problem is the unknowledge of the boundary position for each value of time.

The one-dimensional Stefan Problem consists of

$$\frac{\partial}{-\partial x}\left(k\left(x\right)\frac{\partial u}{\partial x}\right) - \rho c \frac{\partial u}{\partial t} = f\left(x, t\right) \qquad\qquad (4.1.a)$$

$$u\left(0, t\right) = \Gamma \frac{\partial u}{\partial x}\left(0, t\right) + \alpha\left(0, t\right) \qquad\qquad (4.1.b)$$

$$\mu\left(s\left(t\right), t\right) = 0 \qquad\qquad (4.1.c)$$

$$k\left(s\left(t\right)\right)\frac{\partial u}{\partial x}\left(s\left(t\right), t\right) = -\lambda \rho \frac{\partial s}{\partial t} + \mu\left(s\left(t\right), t\right) \qquad (4.1.d)$$

μ is a given boundary source which depends on the position of the free boundary. $k\left(t\right)$ and all other data functions are assumed to be piece-wise continuous in this model. At points of discontinuity of the data one-sided limits are to be taken.

We solve the Stefan Problem combining the OSM and the Method of Lines like was done for the Sweep Method in [4].

The first step is to approximate the parabolic equation by a sequence of elliptic problems at successive time steps by discretizing $\frac{\partial u}{\partial t}$. Let t_n denotes the nth time level and set $\Delta t = t_{n+1} - t_n$. Then $(4.1.a) - (4.1.d)$ can be approximated by

$$Lu = k\left(u'\right)' - \rho c \frac{u}{\Delta t} = -\rho c \frac{u_{n-1}}{\Delta t} + f\left(x, t_n\right) \qquad\qquad (4.2.a)$$

$$u\left(0\right) = \Gamma u'\left(0\right) + \alpha\left(t_n\right) \qquad\qquad (4.2.b)$$

$$u\left(s\right) = 0 \qquad\qquad (4.2.c)$$

$$k\left(s\right)u'\left(s\right) = -\lambda \rho \frac{s - s_{n-1}}{\partial t} + \mu\left(s, t_n\right) \qquad\qquad (4.2.d)$$

and now we applied the following algorithm for $n = 1, 2, \ldots$:

Step 1. We take s_{n-1} as initial approximations for s_n.

Step 2. We solve the BVPs $(4.2.a) - (4.2.c)$ by the OSM and using the last approximation for s_n in place of it.

Step 3. Solving the scalar equation (4.2.d) for s we find a new approximation for s, and if the imposed convergence condition is not fulfilled, we go back to step 3. The numerical results for $k = \rho = c = \lambda = 1$, $f = \mu = \Gamma = 0$, $\alpha = -1$, whose exact solution is

$$u(x,t) = -1 + \frac{\Phi\left(\frac{x}{\sqrt{t}}\right)}{\Phi(0.6202)}, \quad 0 \leq x \leq s(t), \quad \Phi(x) = \frac{2}{\sqrt{\pi}} \int_0^x e^{-t^2} d\tau$$

$$s(t) = 1.2404\sqrt{t}$$

taking $\Delta t = \Delta x = 0.01$ and $t = 0.5$.

4.1.7 Unbounded Interval
We also applied the OSM to solve the Holt's Problem

$$y" - \left(x^2 + R\right)y = 0$$

$$y(0) = \beta \qquad\qquad \lim_{x\to\infty} y(x) = 0$$

where $R = 2m + 1$, $m \in N$. This problem is considered a classical test problem. In this case the OSM is simpler to apply since the system (2.7.a)-(2.7.b) results

$$\sqrt{\varepsilon}z' - q(x)z = \sqrt{\varepsilon}q'(x)y \qquad\qquad (4.3.a)$$

$$\sqrt{\varepsilon}y' - q(x)y = z(x) \qquad\qquad (4.3.b)$$

where $\varepsilon = \frac{1}{R}$ and $q(x) = \sqrt{1 + \frac{x^2}{R}}$. So taking the boundary conditions $z(0) = 0$ and $y(0) = \beta$, we guarantee that the solution of the above system is the solution of our original problem. Thus the algorithm for this case is:

Step 1. Given an approximation for y, we solve the problem (4.3.a).
Step 2. Given z, we solve the problem (4.3.b).
Step 3. If the imposed convergence condition is not fulfilled go back to Step 1.
There are some techniques wich change the second boundary condition for one equivalent in a finite point, [14], but for this problem was enough to take the condition $y(10) = 10^{-30}$.

4.1.8 Orr-Sommerfield Equation[7]
This equation is a typical problem that appears in fluid mechanics. The following eigenvalue problem was solved using the OSM;

$$y^{IV} - 2\alpha y" + \alpha^4 = i\alpha R\left[(U - c)\left(y" - \alpha^2 y\right) - U"y\right], \quad x[0,1]$$

32

$$y'(0) = y''(0) = 0 \qquad \text{and} \qquad y(1) = y'(1) = 0$$

where $\alpha = 1$, $R = 10^{+4}$ (Reynolds number), $U = 1 + x^2$ and $c = c_r + ic_i$ is the eigenvalue.

This example has a turning point for $c_i = 0$ and $c_r \approx 1 - x^2$. The obtained value of c was $(0.237501 + i0.0037378)$.

Here it is necessary to take a non uniform grid given by

$$x_j = w_j \exp\left[\rho\left(w_j^2 - 1\right)\right], \quad j = 1, 2, .., N, \quad \rho = \log R$$

The number of points was $N = 500$ and the relative error was 10^{-7}.

The execution time was relatively longer because of the second iterative process that was necessary to implement for the eigenvalue. We used for that inverse iteration method.

5 Conclusions

The Operator Splitting Method is a suitable technique to solve a large variety of singularly perturbed problems arriving from the physical or engineering problems.

The Quadrature formulae provides a very good tool to solve the linear stiff IVPs.

The Method of Lines gives us the possibility of transformation of the PDEs in a sequence of BVPs, and then to use the (OSM).

The authors are working in some others possibilities of (OSM) and its applications.

Acknowledgments

The authors want to thank to Dr. A. A. Abramov who supplied the original ideas of the OSM.

References

[1] KREISS H. O. , NICHOLS N. K. , Numerical methods for stiff two-point boundary value problems, SIAM JNum. Anal. Vol. 23, No 2, April 1986.

[2] ALVAREZ DIAZ L., RODRIGUEZ SANTIESTEBAN A. , The operator descomposition method for solution of a class of singularly perturbed problems, Serie Tecnica e Investigacion, Vinculos matematicos # 130-1992, Departamento de Matematicas, Facultad de Ciencias, UNAM, Mexico.

[3] HEMKER P.W., Numerical Study of Stiff two-point Boundary Problems, Matematisch Centrum, Amsterdam,1977.

[4] MEYER G. H , Front Traking for the Conductive Stefan Problem with Surface Tension, International Series of Numerical Mathematics, Vol. 86, 1988.

[5] ASCHER U. M. , ROBERT M. M. , RUSSEL R. D., Numerical solution of boundary problems for Ordinary Differential Equations, Englewood Cliffs, New Jerry, Prentice-Hall, Inc. 1988.

[6] FRAILE RODRIGUEZ C. L. ,ALVAREZ L. M., RODRIGUEZ SANTIESTEBAN A., Solución numerica de problemas de contorno de perturbacion singular de tipo oscilatorio, Rev. Investigacion Operacional, Vol. 15, No 2,1994, Cuba.
[7] ALVAREZ L. M., DITKIN V. V., U. S. S. R. Comp. Maths. Math. Phys, Vol. 30, No 4, pp 611-615, 1990.

JOSÉ LUIZ BOLDRINI and MARKO A. ROJAS-MEDAR

Global strong solutions of the equations for the motion of nonhomogeneous incompressible fluids

1 Introduction

In this work we will be concerned with global existence in time of strong solutions of the three dimensional stratified (or nonhomogeneous) Navier-Stokes equations in the case that the external force field does not decay with time. These equations govern the motion of a nonhomogeneous incompressible fluid, obtained as a mixture of miscible incompressible fluids, for instance. Being $\Omega \subset {I\!\!R}^n$, $n = 2$ or 3, a $C^{1,1}$-regular bounded open set, these equations are:

$$\begin{cases} \rho u_t + \rho u.\nabla u - \Delta u - \nabla p = \rho f, \\ \text{div } u = 0, \\ \rho_t + u.\nabla \rho = 0, \quad \text{in } \Omega; \\ u = 0 \ \text{ on } \ \partial\Omega \times (0,\infty), \rho|_{t=0}(x) = \rho_0, \ u|_{t=0}(x) = u_0 \ \text{ in } \ \Omega \,. \end{cases} \tag{1.1}$$

The physical interpretations and the notations are the usual ones in the context of Navier-Stokes equations; the unknowns in the problem are u, ρ and p. We observe that the classical Navier-Stokes equations correspond to the special case where $\rho(x,t) = \rho_0$ is a positive constant; in this case the third equation in (1.1) is automatically satisfied. This case has been much studied (see Ladyszhenskaya [10] and Temam [17] and the references there in.)

Equations (1.1) have been less studied, maybe due to their mixed parabolic-hyperbolic character and stronger nonlinearities. Antonzev and Kazhikov [2], Kazhikov [7], Lions [12], Simon [16] and Kim [9] have studied local and global existence of weak solutions of (1.1). Stronger local and global solution were obtained by Ladyszhenkaya and Solonnikov [11], by linearization and fixed point arguments, and also by Okamoto [13], by evolution operators techniques together with fixed point arguments.The more constructive spectral semi-Galerkin method was used by Salvi [15] to obtain local in time strong solutions and to study conditions for regularity at $t = 0$. This technique was also used by Boldrini and Rojas-Medar [3] to obtain global strong solutions corresponding to a certain class of external forces. However, all these previously known results required some sort of decay in time of the associated external force in order to prove global existence of strong solutions. Ladyszhenkaya and Solonnikov worked in L^q-type spaces, with $(q > n)$, and required exponential decay in time of the (small) $L^q(\Omega)$-norm of the external force. Okamoto worked with L^2-type spaces with f identically zero, and, in order to extend his results to nonzero force fields, an exponential decay in time of the $L^2(\Omega)$-norm of f would be required. Boldrini and Rojas-Medar also worked with L^2-type spaces and required a milder form of decay, that is, $f \in L^2([0,\infty); (L^2(\Omega))^n)$. In all of these previous results, the decay of the external force was an essential ingredient in the arguments. It was used to handle the technical difficulties associated to the interplay between the nonlinearities in the problem, which are stronger than the ones appearing in the classical Navier-Stokes equations, and the fact that an unknown of the problem (ρ) multiplies the external force field f in the first equation in (1.1), creating thus the possibility of unbounded growth of the derivatives of ρ. For this reason, not even in the case in which the external force field is a gradient of a potential (like in the important case of the gravitacional field,) global existence of strong solutions was guaranteed (the usual trick of incorporating a given potencial in the hydrostatic pressure can not be used in this case.)

On the other hand, in the case of the classical Navier-Stokes equations, the decay in time of the force field is not a essential requirement for existence of global strong solutions (at least under other suitable hypotheses; see, for instance, Heywood and Rannacher [7]). Therefore, one also expects to be able to prove global existence of strong solutions of equations (1.1) without imposing any form of decay of the external force. This is indeed true, and it will be prove in this paper.

We will prove global existence by assuming f belonging to $L^\infty(0, \infty; (H^1(\Omega))^n)$ and f_t belonging to $L^\infty(0, \infty; (L^2(\Omega))^n)$ (with small enough norms of the initial condition and external force in the three dimensional case, as in the case of the classical Navier-Stokes equations, and no restriction on values of such norms in the two dimensional case.) We will also require certain other regularity conditions that will be detailed later on in the paper. We also present a sequence of estimates for the (strong) solutions of (1.1) and their spectral approximations. These estimates are important because they are used in an essential way in a paper by Boldrini and Rojas-Medar [4] to obtain uniform in time error bounds for the spectral approximations of (1.1). These estimates are similar to the ones in Heywood [6] in the case of the classical Navier-Stokes, and they are derived under a certain assumption on the stability of the solution being approximated. Thanks to the estimates present here, these uniform in time error bounds are obtained without non realistic assumptions like the ones in Salvi [14], which require a global compatibility condition on the initial data.

We remark that, from the technical point of view, the hyperbolic character of the transport equation in (1.1) and the extra nonlinearities in the problem make the technical arguments more difficult than those used with the classical Navier-Stokes equations. To obtain the required estimates, in our case it will be necessary to prove first certain auxiliary estimates using exponentials as weightning functions.

Finally, we observe that, with the results of this paper, our knowledge about strong solutions of the stratified (nonhomogeneous) Navier-Stokes equations approaches a level similar to that of the classical Navier-Stokes equations.

2 Global Existence in the Case of External Forces without Decay

We start by recalling certain definitions and facts that will be used in the rest of the paper. In general, we will denote the norm in a Banach space B by $||\cdot||_B$; in particular, the norm in $L^p(\Omega)$ will be denoted by $||\cdot||_{L^p(\Omega)}$ (in this paper we assume that Ω is a open bounded set of class $C^{1,1}$ in $\mathbb{R}^n$, with $n = 1$ or 2.) However, to simplify the notation, in the special case $p = 2$, we will denote the L^2-norm simply by $||\cdot||$. We will also work with the usual Sobolev spaces $W^{m,q}(D) = \{f \in L^q(D); ||\partial^\alpha f||_{L^q(D)} < +\infty, (|\alpha| \le m)\}$, where $m = 0, 1, 2, \ldots$, $1 \le q \le \infty$, $D = \Omega$ or $\Omega \times (0, T)$, $0 < T \le +\infty$, with the usual norm. When $q = 2$, we denote $H^m(D) = W^{m,2}(D)$ and $H_0^m(D) = $ closure of $C_0^\infty(\Omega)$ in $H^m(D)$. Being B a Banach-space, we will denote by $L^q(0, T; B)$ the Banach space of the B-valued functions that are L^q-integrable in the sense of Bochner on $[0, T]$. We will also use the space of continuous B-valued functions $C([0, T); B)$.

Being $C_0^\infty(\Omega)^n$ the set of C^∞-functions with compact support in Ω and values in $\mathbb{R}^n$, let $C_{0,\sigma}^\infty(\Omega) = \{v \in C_0^\infty(\Omega)^n; \text{div } v = 0 \text{ in } \Omega\}$, and define $V = $ closure of $C_{0,\sigma}^\infty(\Omega)$ in $H_0^1(\Omega)^n$, and $H = $ closure of $C_{0,\sigma}^\infty(\Omega)$ in $L_0^2(\Omega)^n$. The inner product in H (i.e., the L^2-inner product) will be denoted by $(\cdot, \cdot)$.

Being P the orthogonal projection from $(L^2(\Omega))^n$ onto H obtained by the usual Helmholtz decomposition, the operator $A : H \to H$ given by $A = -P\Delta$ with domain $D(A) = H^2(\Omega)^n \cap V$ is called the Stokes operator. It is well known that A is a positive-definite self-adjoint operator that is characterized by the relation $(Aw, v) = (\nabla w, \nabla v)$ for all $w \in D(A)$, $v \in V$.

We observe that for the regularity properties of the Stokes operator, it is usually assumed that Ω is of class C^3; this being in order to use Cattabriga's results [5]. We use instead the stronger results of Amrouche and Girault [1] which implies, in particular, that when $Au \in (L^2(\Omega))^n$ then $u \in H^2(\Omega)$ and $||u||_{H^2}$ and $||Au||$ are equivalent norms when Ω is of class $C^{1,1}$.

We will denote respectively by φ_k and λ_k $(k \in N)$ the eigenfunctions and eigenvalues the Stokes operator defined on $V \cap (H^2(\Omega))^n$. It is well known that $\{\varphi^k\}_{k=1}^\infty$ form an orthogonal complete system in the spaces H, V and $V \cap (H^2(\Omega))^n$ with their respective usual inner products (u, v), $(\nabla u, \nabla v)$ and (Au, Av). The space generated by the first k eigenfunctions will be denoted $V_k = \text{span } [\varphi_1, \ldots, \varphi_k]$. The orthogonal projection on V_k will be denoted by P_k. We observe that holds the relation $P_k A = A P_k$.

To finish this matter of notations, we remark that, as it is usual in the context of partial differential equations, in the derivations of the estimates, we will denote $c, c_1, \ldots, C, C_1, \ldots$ generic constants depending only on the fixed data of the problem.

The following assumptions on the initial data will hold throughout this paper:

(A.1) The initial value for the density ρ_0 belongs to $C^1(\overline{\Omega})$ and satisfies $0 < \alpha \le \rho_0(x) \le \beta < +\infty$ in Ω.

(A.2) The initial value u_0 belongs to $V \cap (H^2(\Omega))^n$.

Now, we rewrite problem (1.1) as follows: we have to find $\rho \in C^1(\Omega \times [0, T))$ and

$u \in C([0, T); D(A))$, $u_t \in L^\infty(0, T; H) \cap L^2(0, T; V)$ $(0 < T \leq +\infty)$ such that

$$\begin{cases} (\rho u_t, v) + (\rho u. \nabla u, v) + (Au, v) = (\rho f, v) \, , \, \forall v \in H \, , \\ \rho_t + u. \nabla \rho = 0, \text{ ae } (x, t) \in \Omega \times (0, T) \, , \\ u|_{t=0} = u_0 \, , \, \rho|_{t=0} = \rho_0 \, , \text{ ae } x \in \Omega, . \end{cases} \qquad (2.1)$$

Being u_0^k are the projections of u_0 on V_k, the espectral semi-Galerkin approximations for (u, ρ) are defined for each $k \in N$ as the solution $(u^k, \rho^k) \in C^2([0, T]; V_k) \cap C^1(\overline{\Omega} \times [0, T))$ of

$$\begin{cases} (\rho^k u_t^k, v) + (\rho^k u^k. \nabla u^k, v) + (Au^k, v) = (\rho^k f, v), \, \forall v \in V_k; , \\ \rho_t^k + u^k. \nabla \rho^k = 0 \, , \quad \text{for } \forall (x, t) \in \Omega \times (0, T), \\ u(x, 0) = u_0^k(x) \quad \rho(0, x) = \rho_0(x) \, , \quad \forall x \in \Omega. \end{cases} \qquad (2.2)$$

By using these approximations, Salvi [15] proved a local in time existence theorem for (2.1) under assumptions (A.1) and (A.2). His result (Theorem 1 and 2, in Salvi [17]) is the following:

Proposition 2.1. Suppose that $n = 3$, that (A.1) and (A.2) are true and that $f \in L^2(0, T; (H^1(\Omega))^n$, $f_t \in L^2(0, T; (L^2(\Omega))^n)$; $0 < T < +\infty$. Then, there is $0 < T' \leq T$ such that the solution (u, ρ) of (2.1) satisfies $u \in L^\infty(0, T'; V \cap (H^2(\Omega))^n) \cap L^2(0, T'; (H^3(\Omega))^n \cap V)$, $\rho \in C^1(\Omega \times (0, T'))$, $\alpha \leq \rho(x, t) \leq \beta$, for almost all (x, t) in $\Omega \times (0, T')$. This solution is unique.

Remarks:
(i) The statement of Proposition 2.1 that appears in Salvi [15] requires only $f \in (L^2(\Omega \times (0, T)))^n$ and $f_t \in (L^2(\Omega \times (0, T)))^n$. However, this is a mistake generated by the fact that the proof in [15] is done with $f \equiv 0$; following the proof, we see that in order that the result be true when $f \not\equiv 0$, it is necessary to assume the regularity of f that we have stated above. We also observe that it is easily seen that in the proof of Proposition 2.1 the semi-Galerkin approximations exist globally in time (see also Kim [9].)
(ii) The above stated estimate for the density ρ is consequence of the continuity equation (equation (2.1)(ii)) and hypotheses (A1) and (A2). It also holds for the approximations ρ^k. Thus, we have both $0 < \alpha \leq \rho \leq \beta$ and $0 < \alpha \leq \rho^k \leq \beta$. We will use these estimates in the rest of the paper.

Now, we state our first result on global existence of strong solutions.

Theorem 2.1. Suppose that $n = 3$, that (A.1) and (A.2) are true and that $f \in L^\infty(0, \infty; (H^1(\Omega))^n)$, $f_t \in L^\infty(0, \infty; (L^2(\Omega))^n)$. Then, when $\|u_0\|_{H^1(\Omega)}$ and $\|f\|_{L^\infty(0,\infty;(L^2(\Omega))^n)}$ are sufficiently small, the solution (ρ, u) of problem (2.1) exists globally in time and satisfies $u \in C([0, \infty); (H^2(\Omega))^n \cap V)$, $\rho \in C^1(\overline{\Omega} \times [0, T])$ for any finite $T > 0$. Also, for any $\gamma > 0$ there exists some finite positive constants M and C such that

$$\sup_{t \geq 0} \|\nabla u(t)\| = M, \qquad (2.3)$$

38

$$\sup_{t\geq 0} ||u_t|| \leq C \quad \text{and} \quad \sup_{t\geq 0} e^{-\gamma t} \int_0^t e^{\gamma s} ||\nabla u_t(s)||^2 ds \leq C, \tag{2.4}$$

$$\sup_{t\geq 0} ||Au(t)|| \leq C, \tag{2.5}$$

$$\sup_{t\geq 0} e^{-\gamma t} \int_0^t e^{\gamma s} ||u(s)||_{W^{2,6}}^2 ds \leq C. \tag{2.6}$$

Moreover, the same kind of estimates hold uniformly in k for the semi-Galerkin approximations.

Remark: As usual, in this and the other results of this paper, the global existence of strong solutions will follow just by proving the state estimates. Also, we remark that the above estimates should be proved first for the approximations (ρ^k, u^k) and then carried to (ρ, u) in the limit. Since that is a standard procedure and the being all the computations with the approximate solutions exactly the same as those to be done formally with the solution, to simplify the notation, we will work directly with (ρ, u) in the rest of the paper. For heigher order estimates it is important to observe that we are using a spectral basis to build the approximations.

Proof: We start by proving the boundness in time of $||\nabla u(t)||$. ¿From Kim [9], p. 94, we have the following differential inequality

$$\frac{d}{dt}||\nabla u||^2 + c||Au||^2 + c||\rho^{1/2}u_t||^2 \leq c_1||\nabla u||^{10} + c_3 \tag{2.7}$$

where c, c_1 and $c_3 = c\sup_{t\geq 0}||f(t)||$ are positive constants. From the ellipticicity of A, there exists $c_2 > 0$ such that $c_2||\nabla u||^2 \leq c||Au||^2$, and so, the above inequality implies that $\frac{d}{dt}||\nabla u||^2 \leq c_1||\nabla u||^{10} - c_2||\nabla u||^2 + c_3$, with positive constants c_i. Thus, results on differential inequalities can then be applied to conclude that for small enough $||\nabla u_0||$ and c_3 (and, therefore, small enough $\sup_{t\geq 0}||f(t)||$,) there is there is a constant $M > 0$ such that (2.3) holds.

To obtain the other estimates, we will use exponentials as weightning functions in time. As in Heywood and Rannacher [7], we multiply (2.7) by $e^{\gamma t}$, with any $\gamma > 0$, integrate in time from 0 to t, and multiply the resulting inequality by $e^{-\gamma t}$, after several computations, using the result that $||\nabla u(t)||$ is uniformly bounded, we get the auxiliary results that $e^{-\gamma} \int_0^t e^{\gamma s}||u_t||^2 ds$ and $e^{-\gamma t} \int_0^t e^{\gamma s}||Au(s)||^2$. are uniformly bounded in time.

Now, by differentiating (2.1) with respect to t and setting $v = u_t$, after several integrations by parts and estimations using suitable Sobolev and interpolations inequalities, as well as Young's inequalities, we obtain

$$\frac{d}{dt}||\rho^{1/2}u_t||^2 + ||\nabla u_t||^2 \leq C(\beta, M)(||u_t||^2 + ||Au||^2 + ||f_t||^2 + ||\nabla f||^2).$$

We observe that some of the terms appearing in the derivation of the above inequalities are similar to corresponding ones appearing when one handles simarly the classical

39

Navier-Stokes equations. However, there are several other new terms where ρ_t appears (obviously these terms do not show in the case of the classical Navier-Stokes equations;) they are treated by using the continuity equation (2.1)(ii), which is the same as $\rho_t = -\mathrm{div}(\rho u)$, and integration by parts.

Now, we multiply the above inequality by $e^{\gamma t}$, integrate in time from 0 to t, to obtain

$$e^{\gamma t}||\rho^{1/2}u_t||^2 + \int_0^t e^{\gamma s}||\nabla u_t(s)||^2 ds \le \beta ||u_t(0)||^2$$
$$+ c(\beta, M, \gamma) \int_0^t e^{\gamma s}(||u_t(s)||^2 + ||Au(s)||^2) ds + \int_0^t e^{\gamma s}(||f_t(s)||^2 + ||\nabla f(s)||^2) ds. \qquad (2.8)$$

Then, we multiply (2.8) by $e^{\gamma t}$ and use the hypotheses on f and the above auxiliary estimates, together with an estimate for $||u_t(0)||^2$ that can be easly obtained with the help of the equation (2.1)(i) at $t = 0$. The resulting inequality furnishes the estimates stated in (2.4).

The estimates in (2.5) are consequence of the ones in (2.4). It is enough to set $v = Au$ in (2.1) to get $||Au||^2 = (\rho f, Au) - (\rho u_t, Au) - (\rho u \nabla u, Au)$. By estimating the terms in right-hand side with the help of the previous estimates and the hypotheses on f, we obtain (2.5).

Now, we observe that (2.1) is equivalent to $Au = P(\rho(u_t + u\nabla u - f)) = F$ (as observed in the remark after the statement of the theorem, we should actually be working with the approximate solutions. Then the argument should be that (2.2) is equivalent to $Au^k = P_k(\rho^k(u_t^k + u^k \nabla u^k - f)) = F^k$, which holds because we are using a spectral basis; recall that in this case $P_k A = AP_k$.) Since our previous estimates imply that $e^{-\gamma t}\int_0^t e^{\gamma s}||F(s)||^2_{L^6} ds \le C$ (the same holds for F^k), Amrouche and Girault's results [1], imply that (2.6) holds.

To obtain further information on ρ, we observe that, by Sobolev embeddings, the second estimate in (2.6) implies $e^{-\gamma t} \int_0^t e^{\gamma s}||\nabla u(s)||^2_{C(\Omega)} ds \le C$. Also, we observe that, from their proofs, all the previous estimates hold true for $\gamma \ge 0$ when one considers only a finite time interval $[0, T]$, $0 < T < +\infty$ (with the bounding constants obviously depending on T). Thus, on a finite interval $[0, T]$, we can take $\gamma = 0$ in the last result, obtaining $\int_0^t ||\nabla u(s)||^2_{C(\Omega)} ds \le C$. This and an estimate by Ladyzhenskaya and Solonnikov [11], p. 705, for $\nabla \rho$ and ρ_t imply that $\rho \in C^1(\Omega \times [0, T])$ for any finite $T > 0$. $\qquad \square$

Remark: We observe that the same arguments used above can be used to show that the local weak solutions constructed in Kim [9] are in fact global (for small enough data when $n = 3$.)

In what follows we will prove that for the stratified Navier-Stokes equations it is also true that it is not necessary to assume the smallness of the initial data and external force in the two dimensional case in order to obtain global existence of strong solutions.

Theorem 2.2. Suppose that $n = 2$, that (A.1) and (A.2) are true, and that $f \in L^\infty(0, \infty; (H^1(\Omega))^n)$, $f_t \in L^\infty(0, \infty; (L^2(\Omega))^n)$. Then, the solution (ρ, u) of Problem (2.1) exists globally in time and satisfies $u \in C([0, \infty); (H^2(\Omega))^n \cap V)$, $\rho \in C^1(\overline{\Omega} \times [0, T])$ for any finite $T > 0$. Moreover, the estimates given in Theorem 2.1 are true for any $\gamma \geq 0$.

Proof: Working as in Lions [12], p. 65, we have $\dfrac{d}{dt}||\rho^{1/2}u||^2 + ||\nabla u||^2 = (\rho f, u)$. By multiplying the above equation by $e^{\overline{\gamma}t}$, recalling the Poincarè inequality $||u||^2 \leq C_\Omega||\nabla u||^2$ for $u \in (H_0^1(\Omega))^n$, after some computations, we conclude that for $0 < \overline{\gamma} \leq 1/(4\beta C_\Omega)$ it holds that $\dfrac{d}{dt}(e^{\overline{\gamma}t}||\rho^{1/2}u||^2) + \dfrac{1}{2}e^{\overline{\gamma}t}||\nabla u||^2 \leq \dfrac{1}{2}C_\Omega^2\beta^2 e^{\overline{\gamma}t}||f||^2$ Integrating in time and multiplying this inequality by $e^{-\overline{\gamma}t}$, we obtain

$$2||\rho^{1/2}u||^2 + e^{-\overline{\gamma}t}\int_0^t e^{\overline{\gamma}s}||\nabla u(s)||^2 ds \leq 2\beta||u_0||^2 + (C_\Omega^2\beta^2)/\overline{\gamma}||f||_{L^\infty}^2. \qquad (2.9)$$

On the other hand, working as in Simon [16], p. 1115, we obtain that $\dfrac{d}{dt}||\nabla u||^2 \leq c||\nabla u||^4 + c_1$ where c is a positive constant that depends only on β and Ω) and $c_1 = c\sup_{t\geq 0}||f(t)||^2$. Now, we observe that $c\psi^2 + c_1 \leq 2c\psi^2$ for $\psi \geq (c_1/c)^{1/2}$, and denote $\ell^* = \max\{(c_1/c)^{1/2}, 1, ||\nabla u_0||^2\}$, Then, it is possible to prove that the last differential inequality implies that for all $t \geq 0$ we have $||\nabla u(t)||^2 \leq \ell^* e^M$, where $\overline{M} \equiv (c+\overline{\gamma}k)\{2\beta||u_0||^2 + [(C_\Omega^2\beta^2)/\overline{\gamma}]||f||_{L^\infty}^2\} + k$, and k is a positive constant such that $\lg\psi \leq k + k\psi$ for all $\psi \in I\!\!R$. In fact, this is proved by observing that, on the intervals of time where eventually $\psi(t) = ||\nabla u(t)||^2$ is larger than ℓ^*, we have that $\dfrac{d}{dt}\psi \leq c\psi^2$, or, equivalently, $\dfrac{d}{dt}\lg\psi \leq c\psi$. By multiplication by $e^{\overline{\gamma}t}$, using that $\lg\psi \leq k + k\psi$, and then integrating in time, after a mutiplication of the resulting inequality by $e^{-\overline{\gamma}t}$ and the use of (2.9), one obtains the required estimate. The rest of analysis is now done exactly as in the tridimensional case. $\qquad \square$

Remark: Under the hypotheses of either Theorem 2.3 or 2.4, by working in a standard way, it is possible to prove that , being (u, ρ) a solution of (2.1), there exists $p \in L^\infty([0, \infty); H^1(\Omega)/I\!\!R)$ such (u, ρ, p) is solution of (1.1).

3 Global Existence in the Case of Exponentially Decaying External Forces

In this Section we will consider the case that the external force field decays exponentially in time. We will show that the solutions of (2.1) have better regularity than that in the last section; we will be even able to prove an uniform in time estimate for the L^∞-norm of the gradient of the density. In detail, we have the following result:

Theorem 3.1. Suppose that $n = 3$, that (A.1) and (A.2) are true, and that for some constant $\overline{\gamma} > 0$ we have $e^{\overline{\gamma}t}f \in L^\infty(0,\infty;(H^1(\Omega))^n)$, $e^{\overline{\gamma}t}f_t \in L^\infty(0,\infty;(L^2(\Omega))^n)$. Then, if $||u_0||_{H^1(\Omega)}$ and $||e^{\overline{\gamma}t}f||_{L^\infty([0,\infty);(L^2(\Omega))^n)}$ sufficiently small, the solution (ρ,u) of problem (2.1) exists globally in time. Moreover, there is a positive constant $\gamma^* \leq \overline{\gamma}$ such that for any $0 \leq \theta < \gamma^*$, there hold the following estimates

$$\sup_{t\geq 0} e^{\gamma^* t}||\nabla u(t)||^2 < +\infty \tag{3.1}$$

$$\sup_{t\geq 0} e^{\theta t}(||u_t(t)||^2 + ||Au(t)||^2) < +\infty \quad \text{and}$$

$$\sup_{t\geq 0} \int_0^t e^{\theta s}(||\nabla u_t(s)||^2 ds < +\infty \tag{3.2}$$

$$\sup_{t\geq 0} \int_0^t e^{\theta s}||u(s)||^2_{W^{2,6}} ds < +\infty \tag{3.3}$$

$$\sup_{t\geq 0}(||\nabla\rho(t)||_{L^\infty} + ||\rho_t(t)||_{L^\infty}) < +\infty \tag{3.4}$$

$$\sup_{t\geq 0} \sigma(t)(||\nabla u_t(t)||^2) < +\infty \quad \text{and} \quad \sup_{t\geq 0} \int_0^t \sigma(s)||u_{tt}(s)||^2 ds \tag{3.5}$$

$$\sup_{t\geq 0} \int_0^t \sigma(s)||Au_t(s)||^2 ds < +\infty . \tag{3.6}$$

In the last two estimates $\sigma(t) = \min\{1, t\}e^{\theta t}$. The same kind of estimates hold for the semi-Galerkin approximations.

Remark: In the above result, we have grouped the estimates in the order that they will be prove. Since some of them will come at the same time in the argument, we have collect them with the same reference number.

Proof: Multiplying the differential inequality (2.7) by $e^{\gamma t}$, with $0 \leq \gamma \leq \overline{\gamma}$, using again the fact that $||\nabla u|| \leq C||Au||$ and the hypotheses on f, we obtain that $\frac{d}{dt}(e^{\gamma t}||\nabla u||^2) \leq c_1 e^{\gamma t}||\nabla u||^{10} - c_2 e^{\gamma t}||\nabla u||^2 + \gamma e^{\gamma t}||\nabla u||^2 + \overline{c}_3$. Here, c_1 and c_2 are suitable positive constants independent of γ and $\overline{c}_3 = c \sup_{t\geq 0} e^{\gamma t}||f(t)||^2 \leq c \sup_{t\geq 0} e^{\overline{\gamma}t}||f(t)||^2 < +\infty$. Now, choosing γ equal to $\gamma^* = \min\{\overline{\gamma}, c_2/2\}$ and denoting $\psi(t) = e^{\gamma^* t}||\nabla u(t)||^2$, we obtain $\frac{d}{dt}\psi \leq c_1\psi^5 - (c_2/2)\psi + \overline{c}_3$, with initial condition $\psi(0) = ||\nabla u_0||^2$. Now, arguments similar to the ones used in Theorem 2.1 prove (3.1) when $||\nabla u_0||$ and $||e^{\overline{\gamma}t}f(t)||_{L^\infty(0,\infty;(L^2(\Omega))^n)}$ are small enough.

To obtain the other estimates, we again multiply the differential inequality (2.7) by $e^{\theta t}$, with $0 \leq \theta < \gamma^*$ and integrate in time. After some computations, we obtain that

$$e^{\theta t}||\nabla u(t)||^2 + c\int_0^t e^{\theta s}||Au(s)||^2 ds + c\int_0^t e^{\theta s}||\rho^{1/2}u_t||^2 ds$$
$$\leq ||\nabla u_0|| + c_1(\sup_{t\geq 0}||\nabla u||)^8(\sup_{t\geq 0} e^{\gamma t}||\nabla u||^2)\int_0^t e^{(\theta-\gamma^*)s} ds$$
$$+\theta(\sup_{t\geq 0} e^{\gamma^* t}||\nabla u||^2)\int_0^t e^{(\theta-\gamma^*)s} ds$$
$$+c(\sup_{t\geq 0}||f(t)||)(\sup_{t\geq 0} e^{\gamma^* t}||f(t)||)\int_0^t e^{(\theta-\gamma^*)s} ds.$$

Using $\theta < \gamma^*$, (A.1) and the hypotheses on f, together with estimate (3.1), we obtain the following auxiliary results:

$$\sup_{t \geq 0} \int_0^t e^{\theta s}(||Au(s)||^2 ds < +\infty \tag{3.7}$$

$$\sup_{t \geq 0} \int_0^t e^{\theta s}||u_t(s)||^2 ds < +\infty. \tag{3.8}$$

Then, we use (2.8) with $\gamma = \theta$, (3.1), (3.7) and (3.8), an estimate for $||u_t(0)||$, which can easily obtained as before, and the exponential decay of f to conclude that $\sup_{t \geq 0} e^{\theta t}||u_t(t)||^2 < +\infty$ and $\sup_{t \geq 0} \int_0^t e^{\theta s}||\nabla u_t(s)||^2 ds < +\infty$. Now, analogously as in Theorem 2.3, we can prove that $\sup_{t \geq 0} e^{\theta s}(||Au(t)||^2 < +\infty$, and, therefore, (3.2) is proved.

We proceed by observing that (2.1) is equivalent to $Au = P(\rho(-u_t - u\nabla u + f)$ and, thus, $A(e^{\theta t}u(t)) = P(e^{\theta t}\rho(-u_t - u\nabla u + f))$, with $0 < \theta < \gamma^*$. Now, we use Amrouch and Girault's results [1], the already proved estimates and our hypotheses on f to conclude that (3.3) holds. Consequently, by Sobolev embedding, we obtain $e^{(\theta/2)t}u \in L^2(0, \infty; (W^{1,\infty}(\Omega)))^n$, and, thus, the formula of Ladyzhenskaya and Solonnikov [11], p. 705, $||\nabla\rho(t)||_{L^\infty} \leq C||\nabla\rho_0||_{L^\infty} e^{\int_0^t ||\nabla u(s)||_{L^\infty} ds}$, furnishes the first estimate in (3.4). The continuity equation (third equation in (1.1)) then gives the second estimate in (3.4).

To obtain higher order estimates, we proceed as follows: we differentiate the first equation in (2.1) with respect to t and set $v = u_{tt}$. After some computations, with the help again of the integration by parts and suitable estimates of the resulting terms, by multiplying the resulting inequality by $\sigma(t) = e^{\theta t}\min\{1, t\}$ and integrating in time from $\varepsilon > 0$ to t, we get

$$\sigma(t)||\nabla u_t(t)||^2 + \alpha \int_\varepsilon^t \sigma(s)||u_{tt}(s)||^2 ds \tag{3.9}$$

$$\leq \sigma(\varepsilon)||\nabla u_t(\varepsilon)||^2 + C \int_\varepsilon^t \sigma(t)\{||f||^2 + ||f_t||^2 + ||\nabla u||^2 + ||\nabla u_t||^2 + ||u_t||^2\} ds.$$

Letting ε approach zero along a suitable subsequence, as in Heywood and Rannacher [7], with the help of our hypotheses on f and our previous estimates, we obtain (3.5)

Finally, the estimate (3.6) follows easily from the previous ones by using $v = Au_t$ in equation (2.1). $\quad\square$

In the two dimensional case we have a stronger result:

Theorem 3.2. Suppose that $n = 2$, that (A.1) and (A.2) are true, and that for some constant $\overline{\gamma} > 0$, $e^{\overline{\gamma} t}f \in L^\infty([0, \infty); (H^1(\Omega))^n)$, $e^{\overline{\gamma} t}f_t \in L^\infty([0, \infty); (L^2(\Omega))^n)$. Then, the solution (ρ, u) of problem (2.1) exists globally in time. Moreover, the estimates in Theorem 3.1 hold true for any $0 \leq \theta < \overline{\gamma}$.

Proof: Again working as in Simon [16], p. 1115, in the two-dimensional case, we obtain $\dfrac{d}{dt}||\nabla u||^2 \leq C||\nabla u||^4 + C||f||^2$ with C a positive constant. Multiplying this inequality by $e^{\bar{\gamma} t}$ and denoting $\psi(t) = e^{\bar{\gamma} t}||\nabla u(t)||^2$ and $c_3 = \underset{t \geq 0}{c \sup} e^{\bar{\gamma} t}||f(t)||^2$, we can rewrite this last inequality as $\dfrac{d\psi}{dt} \leq c_1 \psi^2 + c_3$, which can be analysed exactly as in Theorem 2.3, furnishing thus that $\underset{t \geq 0}{\sup} \, e^{\bar{\gamma} t}||\nabla u(t)||^2 = \underset{t \geq 0}{\sup} \, \psi(t) < +\infty$. The rest of the analysis is completely analogous to the one in the previous theorem. $\qquad\square$

References

[1] AMROUCHE, C AND GIRAULT, V (1991), On the existence and regularity of the solution of Stokes problem in arbitrary dimension, Proc. Japan Acad., 67, Ser. A, 171-175.

[2] ANTONZEV, S N AND KAZHIKOV, A V (1973), Mathematical Study of Flows of Non Homogeneous Fluids, Novossibirks, Lectures at the University, (in Russian).

[3] BOLDRINI, J L AND ROJAS-MEDAR, M (1992), Global solutions to the equations for the motion of stratified incompressible fluids, Mat. Contemp. 3, 1-8.

[4] BOLDRINI, J L AND ROJAS-MEDAR, M (1994), An error estimate uniform in time for spectral semi-Galerkin approximations of the non-homogeneous Navier-Stokes equations, Nume. Func. Anal. and Optimiz., Vol. 15, No. 7 and 8, 775-778.

[5] CATTABRIGA, L (1961), Su un problema al contorno relativo al sistema di equazioni di Stokes, Rend. Sem. Mat. Univ. Padova, 31, 308-340.

[6] HEYWOOD, J G (1982), An error estimate uniform in time for spectral Galerkin approximations of the Navier-Stokes problem, Pac. J. Math. 98, 333-345.

[7] HEYWOOD, J G AND RANNACHER, R (1982),Finite element approximation of the nonstationary Navier-Stokes problem I: regularity of solutions and second order error estimates for spatial discretization, SIAM J. Num. Anal. 19 (1982), 275-311.

[8] KAZHIKOV, A V (1974), Solvability of the initial and boundary-value problem for the equations of motion of an inhomogeneous viscous incompressible fluid, Dok. Akad. Nauk. 216 (1974), 1008-1010. English Transl., Soviet Physis. Dokl., 331-332.

[9] KIM, J U (1987), Weak solutions of an initial boundary-value problems for an incompressible viscous fluid with non-negative density, SIAM J. Math. Anal. 18, 89-96.

[10] LADYZHENSKAYA, O A (1969), *The Mathematical Theory of Viscous Incompressible Flow*, Gordon and Breach, Second Revised Edition, New York.

[11] LADYZHENSKAYA, O A AND SOLONNIKOV, V A (1975), Unique solvability of an initial and boundary value problem for viscous incompressible fluids, Zap. Naučn Sem. Leningrado Otdel Math. Inst. Steklov, 52, 1975, pp. 52–109; English Transl., J. Soviet Math., 9, 1978, pp. 697–749.

[12] LIONS, J L (1978), On some questions in boundary value problems of mathematical physics,in *Proceedings of the Symposium on Non Linear Evolution Equations*, Wisconsin-Madison, 1977, M. Crandall (Ed.), Academic Press, 59-84.

[13] OKAMOTO, H (1984), On the equation of nonstationary stratified fluid motion: uniqueness and existence of the solutions, J. Fac. Sci. Univ. Tokyo, Sect. 1A Math. 30, 615-643.

[14] SALVI, R (1989), Error estimates for the spectral Galerkin approximations of the solutions of Navier-Stokes type equation, Glasgow Math. J., 31, pp. 199–211.

[15] SALVI, R (1991), The equations of viscous incompressible nonhomogeneous fluid: on the existence and regularity, J. Australian Math. Soc, Series B - Applied Mathematics, Vol. 33 Part 1, pp. 94-110.

[16] SIMON, J (1990), Nonhomogeneous viscous incompressible fluids: existence of velocity, density, and pressure, SIAM J. Math. Anal., 21, 1093–1117.

[17] TEMAM, R (1979), *Navier-Stokes Equations, Theory and Numerical Analysis*, North-Holland, Amsterdam.

CLAUDE CARASSO and GREGORY PANASENKO

Simulation and formal asymptotic analysis of processes in a catalytic converter

Abstract: The mathematical model of processes in a catalytic converter is proposed in a hypothesis that the velocity of a gas in the tubes of the converter is known. The first term of the formal asymptotic solution is constructed by homogenization of the periodic structure of the converter.

1   Introduction

The aim of this paper is to construct a mathematical model of a catalytic converter and to develop a formal asymptotic analysis of this model in order to deduce a homogenized model of the converter. To this end we apply a formal homogenization proc edure [2], Ch. IV sections 1,6. The homogenized model is much simpler than the initial one and it can be solved by standard numerical methods.

2   Mathematical model of processes in a catalytic converter

A catalytic converter is considered in the present article as a cylindrical domain which consists of a periodic set of parallelepipedal cells. Their lateral areas are parallel to the axis of the cylindrical domain of the converter. Each periodic cell contains the metallic reinforcement layer, the catalyst layer, and a parallelepipedal hole (a gas tube). The linear size of a section of a cell is much less than the diameter of the whole converter. Therefore the geometrical description contains a small parameter $\varepsilon > 0$, which is the ratio of these characteristic sizes. Hot gas enters the tubes of the converter from one side and it goes out from the opposite side. Inside of the tubes the gas in teracts chemically with the catalyst layer in each tube and it heats up all elements of the converter. Evidently, the rate of chemical reaction depends on the temperature. The aim of construction of mathematical model is to simulate the thermal proc esses in the converter.

2.1   Geometrical description of the catalytic converter

Define first the solid part of the catalytic converter (without part occupied by a gas).
Let $\mu \in]0,1[$. Consider a two dimensional domain

$$B_{\varepsilon,\mu} = \cup_{k=-\infty}^{+\infty}\{x \in \mathbb{R}^2; \quad \mid x_1 - k\varepsilon \mid < \frac{\varepsilon\mu}{2} \ or \ \mid x_2 - k\varepsilon \mid < \frac{\varepsilon\mu}{2}\}.$$

It is a cross-section of the infinite periodic set of parallelepipedal cells (with parallelepipedal holes inside) of the catalytic converter , mentioned above. The cross-

47

section of each such a cell is represented by a domain

$$D_{\varepsilon,\mu}^{(i_1,i_2)} = B_{\varepsilon,\mu} \cap (]i_1\varepsilon, (i_1+1)\varepsilon[\times]i_2\varepsilon, (i_2+1)\varepsilon[).$$

The position of this cell is defined by the coordinates $(i_1\varepsilon, i_2\varepsilon)$ of its left down corner.

Let Ω be a two-dimensional domain with the C^∞ smooth boundary $\partial\Omega$, and let $D_{\varepsilon,\mu}^\Omega$ be the union of all such cross-sections of cells $D_{\varepsilon,\mu}^{(i_1,i_2)}$ which have a nonempty intersection with $\partial\Omega$, i.e. of "boundary cells". Denote $\Omega' = \Omega \backslash \bar{D}_{\varepsilon,\mu}^\Omega$. Define the cartesian product $P = (\Omega' \cap B_{\varepsilon,\mu}) \times]0, l[$ as a geometrical model of the catalytic converter (its solid part). Here l is the length of the converter and the intersection $\Omega' \cap B_{\varepsilon,\mu}$ is its cross-section.

Let $\delta \in]0, 1[$ and denote $M = (\Omega' \cap B_{\varepsilon,\mu\delta}) \times]0, l[$ a geometrical model of the metallic reinforcement. (Remind that

$$B_{\varepsilon,\mu\delta} = \cup_{k=-\infty}^{+\infty} \{x \in \mathbb{R}^2; \quad |x_1 - k\varepsilon| < \frac{\varepsilon\mu\delta}{2} \ or \ |x_2 - k\varepsilon| < \frac{\varepsilon\mu\delta}{2}\}.)$$

We denote also

$$C = ((B_{\varepsilon\mu}\backslash B_{\varepsilon,\mu\delta}) \cap \Omega') \times]0, l[$$

the part of the converter occupied by a catalyst. Thus we have : $P = M \cup C$. The domain $G = (\Omega'\backslash B_{\varepsilon,\mu}) \times]0, l[$ is occupied by a gas (exhaust gas).

2.2 Simulation of processes of heat transfer and diffusion in the catalytic converter

The thermal state of the converter is defined by some chemical reactions between the exhaust gas and the catalyst. The chemical composition of the exhaust gas contains inert gas and such components as $O_2, CO, NO, H_2O, C_3H_6, CH_4$ in a small concentration. Evidently these components penetrate in the porous catalyst layer . We suppose that the gas velocity is known in each tube, i.e. it is a given function of time and space variables. Denote u_{g_i}, $i = 1, \ldots, N$ the concentration of the product i in the gas, u_{c_i}, $i = 1 \ldots, N$ its concentration in the catalyst layer. Let T_c, T_g, T_m be the temperature of the catalyst, the gas and the metal reinforcement respectively. All these physical characteristics are functions of the time variable t and the space variable $x = (x_1, x_2, x_3)$. Denote also $\overline{v}^\varepsilon$ the velocity of the gas in the direction x_3. We assume that it is a given rapidl y oscillating function, which is periodic in x_1 and in x_2 with period ε; i.e. $\overline{v}^\varepsilon(x, t) = \overline{v}(\frac{x_1}{\varepsilon}, \frac{x_2}{\varepsilon}, x_3, t)$, where $\overline{v}(\xi_1, \xi_2, x_3, t)$ is a 1-periodic function of ξ_1 and ξ_2, equal to zero in P.

Writing the heat and mass conservation laws for each chemical component, we obtain the system of equations (see [1]):

$$\rho_g c_g \frac{\partial T_g}{\partial t} + \rho_g c_g \overline{v}^\varepsilon \frac{\partial}{\partial x_3} T_g - k_g (\frac{\partial^2 T_g}{\partial x_1^2} + \frac{\partial^2 T_g}{\partial x_2^2}) = 0 \ in \ G \times]0, T[\tag{1}$$

$$\frac{\partial u_{g_i}}{\partial t} + \overline{v}^\varepsilon \frac{\partial}{\partial x_3} u_{g_i} - k_{g_i}\left(\frac{\partial^2 u_{g_i}}{\partial x_1^2} + \frac{\partial^2 u_{g_i}}{\partial x_2^2}\right) = 0, \quad i = 1, \ldots, N \tag{2}$$

for $(x,t) \in G \times]0, T[$, and

$$\rho_c c_c \frac{\partial T_c}{\partial t} - k_c \Delta T_c = f(u_{c_1} \ldots, u_{c_N}, T_c) \tag{3}$$

$$\frac{\partial u_{c_i}}{\partial t} - k_{c_i}\Delta u_{c_i} = g_i(u_{c_1}, \ldots, u_{c_N}, T_c)\} i = 1, \ldots, N \tag{4}$$

for $(x,t) \in C \times]0, T[$, and

$$\rho_m c_m \frac{\partial T_m}{\partial t} - k_m \Delta T_m = 0 \tag{5}$$

for $(x,t) \in M \times]0, T[$.

Here ρ_g, ρ_c, and ρ_m are respectively, the densities of the gas, of the catalyst and of the metal, c_g, c_c, and c_m are respectively, the calorific capacities of the gas, of the catalyst and of the metal, k_g, k_c, and k_m are respectively, the thermoconductivities of the gas, of the catalyst and of the metal, k_{g_i} and k_i $(i = 1, \ldots, N)$ are respectively the coefficients of diffusion of the product number i in the gas and the catalyst, the right hand sides $f(u_{c_1} \ldots, u_{c_N}, T_c)$ and $g_i(u_{c_1} \ldots, u_{c_N}, T_c)$ are some given functions of the concentrations and the temperature. For example in [1] they are proposed in a following form:

$$g_i(u_{c_1} \ldots, u_{c_N}, T_c) = -\sum_{j=1}^{N} \nu_{ij} R_j$$

with

$$R_j = \alpha^* u_{c_j}\left(\sum_{l=1}^{N} \lambda_{jl} u_{c_l}\right) e^{-\frac{E_j}{T_c}}(T_c K_1 K_2 \ldots K_N)^{-1}),$$

K_m of the form

$$K_m = (1 + \alpha_1^m u_{c_1} + \ldots + \alpha_N^m u_{c_N})^{\beta_m},$$

where ν_{ij} are the stochiometric coefficients of reactions, λ_{jl}, $\alpha_1^m, \alpha_2^m, \ldots, \alpha_N^m$ and β_m are some known constants related to a rate of reaction.

The function f has a form

$$f(u_{c_1}, \ldots, u_{c_N}, T_c) = \sum_{j=1}^{N} R_j J_j^*$$

where J_j^* are proportional to a heat of reaction , i.e. to the liberated heat of reaction j. In the equation (2) we neglected the diffusion effects in x_3 direction in comparison with transport effects. This allows us to formulate the boundary condition for u_g at the entrance of the converter only.

The interface conditions between the gas and the catalyst are:

$$\left.\begin{array}{ll} k_c\frac{\partial T_c}{\partial n} = k_g\frac{\partial T_g}{\partial n}, & T_g = T_c \\ k_{c_i}\frac{\partial u_{c_i}}{\partial n} = k_{g_i}\frac{\partial u_{g_i}}{\partial n}, & u_{c_i} = u_{g_i} \end{array}\right\} \text{ on } (\partial B_{\varepsilon,\mu}\cap\Omega')\times]0,l[$$

The same conditions are posed on the interface between the catalyst and the metal

$$\left.\begin{array}{ll} k_m\frac{\partial T_m}{\partial n} = k_c\frac{\partial T_c}{\partial n}, & T_m = T_c \\ \frac{\partial u_{c_i}}{\partial n} = 0 & \end{array}\right\} \text{ on } (\partial B_{\varepsilon,\mu\delta}\cap\Omega')\times]0,l[$$

The last condition simulates a non-penetrating of the chemical products to the metallic part of the converter. Here $\partial/\partial n$ is a normal derivative.

The initial state of the converter is assumed to be known, i.e. for $t = 0$ and $x \in \Omega'\times]0,l[$ the values of T_g, T_c, T_m, u_{g_i}, and u_{c_i}, $i = 1,\ldots,N$ are given functions of the space variable.

The boundary conditions are as follows: at the entrance of the converter for $x \in \Omega' \times \{0\}$, the values of T_g, T_c, T_m, u_{g_i}, $i = 1,\ldots,N$ are given for $t > 0$ and $\frac{\partial u_{c_i}}{\partial x_3} = 0$, $i = 1,\ldots,N$; at the exit of the converter for $x \in \Omega' \times \{l\}$, the values of T_g, T_c, T_m are given for $t > 0$ and $\frac{\partial u_{c_i}}{\partial x_3} = 0$, $i = 1,\ldots,N$. We suppose here for simplicity that the variation of the temperature at the entrance and at the exit is not too large and it does not depend of rapid variables.

We suppose that the temperature of the lateral boundary $\partial\Omega'\times[0,l]$ of the converter is also given function independent of ε: $T_m(x,t) = T_{m_d}(x,t)$.

This model can be rewritten in a vector form. To this end we define the following piecewise constant functions:

$$c^\varepsilon(x) = \left\{ \begin{array}{lll} \rho_m c_m & \text{if} & x \in M \\ \rho_g c_g & \text{if} & x \in G \\ \rho_c c_c & \text{if} & x \in C, \end{array}\right. \quad d^\varepsilon(x,t) = \left\{ \begin{array}{lll} \overline{v}^\varepsilon \rho_g c_g & \text{if} & x \in G \\ 0 & \text{if} & x \in M_\cup C \end{array}\right.$$

$$k^\varepsilon(x) = \left\{ \begin{array}{lll} k_m & \text{if} & x \in M \\ k_g & \text{if} & x \in G \\ k_c & \text{if} & x \in C \end{array}\right. \quad k_s^\varepsilon(x) = \left\{ \begin{array}{lll} k_m & \text{if} & x \in M \\ k_c & \text{if} & x \in C \\ 0 & \text{if} & x \in G \end{array}\right.$$

we define also

$$T(x,t) = \left\{ \begin{array}{lll} T_m & \text{if} & x \in M \\ T_g & \text{if} & x \in G \\ T_c & \text{if} & x \in C, \end{array}\right. \quad u_i(x,t) = \left\{ \begin{array}{lll} u_{g_i} & \text{if} & x \in G \\ u_{c_i} & \text{if} & x \in C \end{array}\right.$$

Consider the vector of concentrations of N chemical compounds $u = (u_1, ..., u_N)^T$; here T is a transposition symbol. Then we define

50

$$f^\varepsilon(x, u, T) = \begin{cases} f(u, T) & \text{if} \quad x \in C \\ 0 & \text{if} \quad x \in G \cup M, \end{cases}$$

$$g_i^\varepsilon(x, u, T) = \begin{cases} g_i(u, T) & \text{if} \quad x \in C \\ 0 & \text{if} \quad x \in G \cup M \end{cases}$$

and we denote $g^\varepsilon = (g_1, ..., g_N)^T$.

We define the matrix valued function

$$K^\varepsilon(x) = \begin{pmatrix} k_{g_1} & 0 & ... & 0 \\ 0 & k_{g_2} & ... & 0 \\ ... & ... & ... & ... \\ 0 & 0 & ... & k_{g_N} \end{pmatrix}$$

if $x \in G$ and

$$K^\varepsilon(x) = \begin{pmatrix} k_{c_1} & 0 & ... & 0 \\ 0 & k_{c_2} & ... & 0 \\ ... & ... & ... & ... \\ 0 & 0 & ... & k_{c_N} \end{pmatrix}$$

if $x \in C$ and we define also the matrix valued function

$$K_s^\varepsilon(x) = \begin{pmatrix} k_{c_1} & 0 & ... & 0 \\ 0 & k_{c_2} & ... & 0 \\ ... & ... & ... & ... \\ 0 & 0 & ... & k_{c_N} \end{pmatrix}$$

if $x \in C$ and zero matrix if $x \in G$.

Then we can rewrite the equations (1) - (5) in a vector form:

$$c^\varepsilon(x)\frac{\partial T}{\partial t} + d^\varepsilon(x, t)\frac{\partial T}{\partial x_3} - \frac{\partial}{\partial x_3}(k_s^\varepsilon(x)\frac{\partial T}{\partial x_3}) - \sum_{i=1}^{2} \frac{\partial}{\partial x_i}(k^\varepsilon(x)\frac{\partial T}{\partial x_i}) = f^\varepsilon(x, u, T) \qquad (6)$$

for $x \in \Omega' \times]0, l[\times]0, T^*[$,

$$\frac{\partial u}{\partial t} + \overline{v}^\varepsilon(x, t)\frac{\partial u}{\partial x_3} - \frac{\partial}{\partial x_3}(K_s^\varepsilon \frac{\partial u}{\partial x_3}) - \sum_{i=1}^{2} \frac{\partial}{\partial x_i}(K^\varepsilon(x)\frac{\partial u}{\partial x_i}) = g^\varepsilon(x, u, T) \qquad (7)$$

for $x \in (\Omega' \backslash B_{\varepsilon, \mu\delta}) \times]0, l[\times]0, T^*[$.

The boundary conditions are

$$T(x, t) = T_\alpha(x, t) \text{ if } x \in \Omega' \times \{0\} \cup \partial\Omega' \times]0, l[\cup \Omega' \backslash G \times \{l\}; \qquad (8)$$

$$u(x, t) = u_\alpha(x, t) \text{ if } x \in \Omega' \times \{0\} \cup \Omega' \backslash G \times \{l\}; \qquad (9)$$

51

$$\frac{\partial u}{\partial n} = 0 \text{ if } (\partial\Omega' \backslash B_{\varepsilon,\mu\delta}) \times]0, l[\qquad (10)$$

with given smooth T_α and u_α.

The interface conditions for $x \in \partial M \cap \partial C$ and for $x \in \partial C \cap \partial G$ are as usual:

$$\left[k^\varepsilon \frac{\partial T}{\partial n} \right] = 0, \quad [T] = 0, \qquad (11)$$

and for $x \in \partial C \cap \partial G$:

$$\left[K^\varepsilon \frac{\partial u}{\partial n} \right] = 0, \quad [u] = 0. \qquad (12)$$

The initial conditions are

$$T(x,0) = T_0(x), \quad u(x,0) = u_0(x) \qquad (13)$$

with given smooth T_0 and u_0.

The coefficients of the equations (1)-(5) could be represented in a form:

$$c^\varepsilon(x) = c(\frac{\overline{x}}{\varepsilon}), \quad d^\varepsilon(x) = d(\frac{\overline{x}}{\varepsilon}, x_3, t), \quad k^\varepsilon(x) = k(\frac{\overline{x}}{\varepsilon})$$

$$f^\varepsilon(x, u, T) = f(\frac{\overline{x}}{\varepsilon}, u, T), \quad K^\varepsilon(x) = K(\frac{\overline{x}}{\varepsilon})$$

$$g^\varepsilon(x, u, T) = g(\frac{\overline{x}}{\varepsilon}, u, T)$$

where $\overline{x} = (x_1, x_2)$ and the functions c, d, k, f, K, g do not depend on ε and are defined by the relations

$$c(\xi) = \begin{cases} \rho_m c_m & \text{if} \quad \xi \in M_0 \\ \rho_g c_g & \text{if} \quad \xi \in G_0 \\ \rho_c c_c & \text{if} \quad \xi \in C_0, \end{cases}$$

$$d(\xi, x_3, t) = \begin{cases} \overline{v}(\xi, x_3, t)\rho_g c_g & \text{if} \quad \xi \in G_0 \\ 0 & \text{if} \quad \xi \in M_0 \cup C_0, \end{cases}$$

$$k(\xi) = \begin{cases} k_m & \text{if} \quad \xi \in M_0 \\ k_g & \text{if} \quad \xi \in G_0 \\ k_c & \text{if} \quad \xi \in C_0, \end{cases} \quad k_s(\xi) = \begin{cases} k_m & \text{if} \quad \xi \in M_0 \\ k_c & \text{if} \quad \xi \in C_0 \\ 0 & \text{if} \quad \xi \in G_0, \end{cases}$$

$$f(\xi, u, T) = \begin{cases} f(u, T) & \text{if} \quad \xi \in C_0 \\ 0 & \text{if} \quad \xi \in G_0 \cup M_0, \end{cases}$$

52

$$g(\xi, u, T) = \begin{cases} g(u, T) & \text{if} \quad \xi \in C_0 \\ 0 & \text{if} \quad \xi \in G_0 \cup M_0, \end{cases}$$

$$K(\xi) = \begin{pmatrix} k_{g_1} & 0 & \dots & 0 \\ 0 & k_{g_2} & \dots & 0 \\ \dots & \dots & \dots & \dots \\ 0 & 0 & \dots & k_{g_N} \end{pmatrix}$$

if $\xi \in G_0$ and

$$K(\xi) = \begin{pmatrix} k_{c_1} & 0 & \dots & 0 \\ 0 & k_{c_2} & \dots & 0 \\ \dots & \dots & \dots & \dots \\ 0 & 0 & \dots & k_{c_N} \end{pmatrix}$$

if $\xi \in C_0$;

$$K_s(\xi) = \begin{pmatrix} k_{c_1} & 0 & \dots & 0 \\ 0 & k_{c_2} & \dots & 0 \\ \dots & \dots & \dots & \dots \\ 0 & 0 & \dots & k_{c_N} \end{pmatrix}$$

if $\xi \in C_0$ and it is a zero matrix if $\xi \in G_0$; here M_0, C_0, G_0 are two-dimensional 1-periodic sets corresponding to a cross-section of homothetically dilatated in $1/\varepsilon$ times sets M, C and G_0 respectively:

$$M_0 = B_{1,\mu\delta}, \quad C_0 = (B_{1,\mu} \backslash B_{1,\mu\delta}), \quad G_0 = I\!\!R^2 \backslash B_{1,\mu}.$$

Therefore the equations could be rewritten in a form:

$$c(\frac{\overline{x}}{\varepsilon})\frac{\partial T}{\partial t} + d(\frac{\overline{x}}{\varepsilon}, x_3, t)\frac{\partial T}{\partial x_3} - \frac{\partial}{\partial x_3}(k_s(\frac{\overline{x}}{\varepsilon})\frac{\partial T}{\partial x_3}) - \sum_{i=1}^{2} \frac{\partial}{\partial x_i}(k(\frac{\overline{x}}{\varepsilon})\frac{\partial T}{\partial x_i}) = f(\frac{\overline{x}}{\varepsilon}, u, T) \quad (14)$$

for $x \in \Omega' \times]0, l[\times]0, T^*[$,

$$\frac{\partial u}{\partial t} + \overline{v}(\frac{\overline{x}}{\varepsilon}, x_3, t)\frac{\partial u}{\partial x_3} - \frac{\partial}{\partial x_3}(K_s(\frac{\overline{x}}{\varepsilon})\frac{\partial u}{\partial x_3}) - \sum_{i=1}^{2} \frac{\partial}{\partial x_i}(K(\frac{\overline{x}}{\varepsilon})\frac{\partial u}{\partial x_i}) = g(\frac{\overline{x}}{\varepsilon}, u, T) \quad (15)$$

for $x \in (\Omega' \backslash B_{\varepsilon,\mu\delta}) \times]0, l[\times]0, T^*[$.

Thus these equations (14),(15) with the mentioned above boundary conditions (11),(12), interface conditions and initial conditions (13) form the mathematical model of processes in the catalytic converter.

3   Formal asymptotic derivation of a homogenized model

Following the standard homogenization procedure described in [2] we introduce the rapid variables $\xi_1 = x_1/\varepsilon$, $\xi_2 = x_2/\varepsilon$ and we seek the asymptotic solution of the problem (11)-(15) in a form of a sum of the macroscopic (homogenized) solution and the first and second order rapidly oscillating correctors; i.e. the asymptotic approximation $T^{(2)}$, $u^{(2)}$ for the solution T, u will be constructed in a form of a truncated expansion

$$T^{(2)}(x,t) = T_0(x, \frac{\overline{x}}{\varepsilon}, t) + \varepsilon T_1(x, \frac{\overline{x}}{\varepsilon}, t) + \varepsilon^2 T_2(x, \frac{\overline{x}}{\varepsilon}, t) \qquad (16)$$

and

$$u_{(2)}(x,t) = u_0(x, \frac{\overline{x}}{\varepsilon}, t) + \varepsilon u_1(x, \frac{\overline{x}}{\varepsilon}, t) + \varepsilon^2 u_2(x, \frac{\overline{x}}{\varepsilon}, t) \qquad (17)$$

with 1-periodic in ξ functions $T_i(x, \xi, t)$ and $u_i(x, \xi, t_i)$.

Substituting these expressions to the equations (14), (15) and denoting the operators

$$\tilde{L}_{\alpha\beta} = \sum_{l=1}^{2} \frac{\partial}{\partial \alpha_l}(k \frac{\partial}{\partial \beta_l})$$

and

$$L_{\alpha\beta} = \sum_{l=1}^{2} \frac{\partial}{\partial \alpha_2}(K \frac{\partial}{\partial \beta_l})$$

we obtain :

$$\{\sum_{i=0}^{2} (\varepsilon^i c(\xi)\frac{\partial T_i}{\partial t} + \varepsilon^i d(\xi, x_3, t)\frac{\partial T_i}{\partial x_3} - \varepsilon^i (k_s(\xi)\frac{\partial T_i}{\partial x_3})$$

$$-\varepsilon^i \tilde{L}_{xx} T_i - \varepsilon^{i-1}(\tilde{L}_{\xi x} T_i + \tilde{L}_{x\xi} T_i) - \varepsilon^{i-2} \tilde{L}_{\xi\xi} T_i)$$

$$-f(\xi, \sum_{i=0}^{2} \varepsilon^i u_i, \sum_{i=0}^{2} \varepsilon^i T_i)\}|_{\xi=\frac{\overline{x}}{\varepsilon}} = 0$$

and

$$\{\sum_{i=0}^{2} (\varepsilon^i \frac{\partial u_i}{\partial t} + \varepsilon^i \overline{v}(\xi, x_3, t)\frac{\partial u_i}{\partial x_3} - \varepsilon^i \frac{\partial}{\partial x_3}(K_s(\xi)\frac{\partial u_i}{\partial x_3})$$

$$\varepsilon^i L_{xx} u_i - \varepsilon^{i-1}(L_{\xi x} u_i + L_{x\xi} u_i) - \varepsilon^{i-2} L_{\xi\xi} u_i$$

$$-g(\xi, \sum_{i=0}^{2} \varepsilon^i u_i, \sum_{i=0}^{2} \varepsilon^i T_i))\}|_{\xi=\frac{\overline{x}}{\varepsilon}} = 0.$$

Applying Taylor formula to f and g and grouping the terms of the same order in ε we obtain

54

$$\{-\varepsilon^{-2}\tilde{L}_{\xi\xi}T_0 - \varepsilon^{-1}(\tilde{L}_{\xi x}T_0 + \tilde{L}_{x\xi}T_0 + \tilde{L}_{\xi\xi}T_1)$$

$$\varepsilon^0(c(\xi)\frac{\partial T_0}{\partial t} + d(\xi, x_3, t)\frac{\partial T_0}{\partial x_3} - \frac{\partial}{\partial x_3}(K_s(\xi, x_3)\frac{\partial T_0}{\partial x_3})$$

$$-\tilde{L}_{xx}T_0 - f(\xi, u_0, T_0) - \tilde{L}_{\xi x}T_1 - \tilde{L}_{x\xi}T_1 - \tilde{L}_{\xi\xi}T_2)\}|_{\xi=\frac{x}{\varepsilon}} + O(\varepsilon) = 0,$$

$$\{-\varepsilon^{-2}L_{\xi\xi}u_0 - \varepsilon^{-1}(L_{\xi x}u_0 + L_{x\xi}u_0 + L_{\xi\xi}u_1)$$

$$\varepsilon^0(\frac{\partial u_0}{\partial t} + \bar{v}(\xi, x_3, t)\frac{\partial u_0}{\partial x_3} - \frac{\partial}{\partial x_3}(K_s(\xi, x_3)\frac{\partial u_0}{\partial x_3})$$

$$-L_{xx}u_0 - g(\xi, u_0, T_0) - L_{\xi x}u_1 - L_{x\xi}u_1 - L_{\xi\xi}u_2)\}|_{\xi=\frac{x}{\varepsilon}} + O(\varepsilon) = 0,$$

Now the equations should be satisfied up to terms of order $O(\varepsilon)$.
The vanishing of the coefficients of $\varepsilon^{-2}, \varepsilon^{-1}$ and ε^0 derives the equations :

$$L_{\xi\xi}u_0 = 0 \quad \text{and} \quad \tilde{L}_{\xi\xi}T_0 = 0 \tag{18}$$

$$\begin{cases} L_{\xi x}u_0 + L_{x\xi}u_0 + L_{\xi\xi}u_1 = 0 \\ \tilde{L}_{\xi x}T_0 + \tilde{L}_{x\xi}T_0 + \tilde{L}_{\xi\xi}T_1 = 0 \end{cases} \tag{19}$$

$$\begin{cases} \frac{\partial u_0}{\partial t} + \bar{v}(\xi, x_3, t)\frac{\partial u_0}{\partial x_3} - \frac{\partial}{\partial x_3}(K_s(\xi)\frac{\partial u_0}{\partial x_3}) \\ -L_{xx}u_0 - g(\xi, u_0, T_0) - L_{\xi x}u_1 - L_{x\xi}u_1 - L_{\xi\xi}u_2 = 0, \\ c(\xi)\frac{\partial T_0}{\partial t} + d(\xi, x_3, t)\frac{\partial T_0}{\partial x_3} - \frac{\partial}{\partial x_3}(k_s(\xi)\frac{\partial T_0}{\partial x_3}) \\ -\tilde{L}_{xx}T_0 - f(\xi, u_0, T_0) - \tilde{L}_{\xi x}T_1 - \tilde{L}_{x\xi}T_1 - \tilde{L}_{\xi\xi}T_2 = 0. \end{cases} \tag{20}$$

The boundary conditions $\frac{\partial u}{\partial n} = 0$ on $\partial M \times]0, l[$ could be developed as

$$\frac{\partial u^{(2)}}{\partial n} = \varepsilon^{-1}\frac{\partial u_0}{\partial n_\xi} + (\frac{\partial u_0}{\partial n_x} + \frac{\partial u_1}{\partial n_\xi}) + \varepsilon(\frac{\partial u_1}{\partial n_x} + \frac{\partial u_2}{\partial n_\xi}) + \varepsilon^2\frac{\partial u_2}{\partial n_x};$$

here $\frac{\partial}{\partial n_x}$ and $\frac{\partial}{\partial n_\xi}$ are the normal derivatives in the variables x and ξ respectively; this asymptotic development generates the boundary conditions:

$$\frac{\partial u_0}{\partial n_\xi} = 0, \quad \frac{\partial u_0}{\partial n_x} + \frac{\partial u_1}{\partial n_\xi} = 0, \quad \frac{\partial u_1}{\partial n_x} + \frac{\partial u_2}{\partial n_\xi} = 0.$$

In the same manner the interface conditions (11), (12) can be transformed into the natural interface conditions for the equations (18)-(20), corresponding to the variational formulation of these problems.

The functions u_0 and T_0 are 1-periodic in ξ solutions of the homogeneous equations (18) and therefore they are independent of the rapid variables ξ :

$$u_0(x,\xi,t) = v(x,t), \quad T_0(x,\xi,t) = w(x,t).$$

Therefore the equations (19) can be presented in a form

$$L_{\xi\xi}u_1 = -L_{\xi x}v, \quad \tilde{L}_{\xi\xi}T_1 = -\tilde{L}_{\xi x}w,$$

i.e.

$$\sum_{l=1}^{2}\frac{\partial}{\partial\xi_l}(K(\xi)\frac{\partial}{\partial\xi_l}u_1) = -\sum_{l=1}^{2}\frac{\partial}{\partial\xi_l}(K(\xi)\frac{\partial v}{\partial x_l})$$

and

$$\sum_{l=1}^{2}\frac{\partial}{\partial\xi_l}(k(\xi)\frac{\partial}{\partial\xi_l}T_1) = -\sum_{l=1}^{2}\frac{\partial}{\partial\xi_l}(k(\xi)\frac{\partial w}{\partial x_l})$$

Separating x and ξ variables we seek the solution in a form of a linear combination of the derivatives of v and w with rapidly oscillating coefficients:

$$u_1(x,\xi,t) = N_1(\xi)\frac{\partial v}{\partial\xi_1}(x,t) + N_2(\xi)\frac{\partial v}{\partial x_2}(x,t)$$

and

$$T_1(x,\xi,t) = M_1(\xi)\frac{\partial w}{\partial x_1}(x,t) + M_2(\xi)\frac{\partial w}{\partial x_2}(x,t)$$

where N_1 and N_2 are $N \times N$ matrices and M_1 and M_2 are fonctions. They are the 1-periodic weak solutions of the following cell problems:

$$\sum_{l=1}^{2}\frac{\partial}{\partial\xi_l}(K(\xi)\frac{\partial}{\partial\xi_l}N_k(\xi)) = -\frac{\partial}{\partial\xi_q}K(\xi), \quad q = 1,2 \tag{21}$$

in $G_0 \cup C_0$ with boundary conditions on $\partial(G_0 \cup C_0)$

$$\sum_{l=1}^{2}n_l(K(\xi)\frac{\partial}{\partial\xi_l}N_q(\xi)) = -n_qK(\xi), \tag{22}$$

and

$$\sum_{l=1}^{2}\frac{\partial}{\partial\xi_l}(k(\xi)\frac{\partial}{\partial\xi_l}M_q(\xi)) = -\frac{\partial k(\xi)}{\partial\xi_q}, \quad q = 1,2. \tag{23}$$

We impose the condition of a vanishing average

$$< M_q >= 0, \quad < N_q >_s = 0, \tag{24}$$

where

$$< . >= \int_{\mathcal{D}} d\xi, \quad < . >_s= \int_{\mathcal{D} \cap B_{1,\mu}} d\xi, \quad \mathcal{D} =]-1/2, 1/2[^2.$$

The solution of the problem (21),(22),(24) is a linear function $N_q = -\xi_q I$, where I is an identity matrix. The equation (23) with the periodicity condition and with the condition (24) has a unique solution which can be found by means of n umerical analysis [3].

Rewrite the second equation (20) in a form

$$\tilde{L}_{\xi\xi} T_2 = -\tilde{L}_{\xi x} T_1 - \sum_{l=1}^{2} \sum_{q=1}^{2} \frac{\partial}{\partial x_l} (k(\xi) \frac{\partial M_q(\xi)}{\partial \xi_l} \frac{\partial w}{\partial x_q})$$

$$-\tilde{L}_{xx} w - \frac{\partial}{\partial x_3} (k_s(\xi, x_3) \frac{\partial w}{\partial x_3}) + c(\xi) \frac{\partial w}{\partial t} + d(\xi, x_3, t) \frac{\partial w}{\partial x_3} - f(\xi, u_0, w). \qquad (25)$$

The condition of existence of 1-periodic solution of this equation is the condition of vanishing of the average of the right hand side function (cf. [2], p.112), i.e. the relation hold true for all $x \in \Omega \times]0, l[$ and $t \in]0, T^*[$

$$< c > \frac{\partial w}{\partial t} + < d > \frac{\partial w}{\partial x_3} - \frac{\partial}{\partial x_3} (< k_s > \frac{\partial w}{\partial x_3})$$

$$- \sum_{l=1}^{2} \sum_{q=1}^{2} \frac{\partial}{\partial x_l} (\hat{a}_{lq} \frac{\partial w}{\partial x_q}) - < f(., v, w) >= 0, \qquad (26)$$

where $\hat{a}_{lq} =< k \frac{\partial(M_q + \xi_q)}{\partial \xi_l} >$ are called the effective thermoconductivities in the plane $x_1 \ O \ x_2$ and they are calculated as the integrals of fluxes for the equation (23) mentioned above. Using parity properties of the solution of cell problem (cf. [2]) we obtain the diagonality of the matrix $\hat{a}$ and using a symmetry of the geometry of the cell we obtain an isotropy of the structure, i.e. $\hat{a} = \hat{a}_{11} I$. Thus there is only one effective thermoconductivity coefficient in the plane $x_1 \ O \ x_2$, which should be determined by means of numerical analysis.

In a same way we rewrite the first equation (20) with the boundary condition $\frac{\partial u_2}{\partial n_\xi} + \frac{\partial u_1}{\partial n_x} = 0$. This Neumann problem for the elliptic equation has a solvability condition, derived from the Ostrogradsky formula:

$$mes \ (\mathcal{D} \cap B_{1,\mu}) \frac{\partial v}{\partial t} + < \bar{v} > \frac{\partial v}{\partial x_3} - \frac{\partial}{\partial x_3} (< K_s > \frac{\partial v}{\partial x_3}) =< g(., v, w) >, \qquad (27)$$

which we impose for all $x \in \Omega \times]0, l[$ and $t \in]0, T^*[$.

If the conditions (26) and (27) are satisfied then the problems (20) have the periodic solutions and therefore the asymptotic approximation $T^{(2)}$, $u^{(2)}$ satisfies the equations (6) and (7) up to $O(\varepsilon)$. The boundary condition (1 0) and interface conditions (11), (12) are satisfied up to $O(\varepsilon^2)$. Now we consider the relations (26) and (27) as the equations for the homogenized (macroscopic) temperature w and homogenized (macroscopic) concentration v. Let w and v satisfy the boundary conditions

$$w(x,t) = T_\alpha(x,t) \text{ if } x \in (\Omega \times \{0\}) \cup (\partial\Omega \times]0, l[) \cup (\Omega \times \{l\}); \tag{28}$$

$$v(x,t) = u_\alpha(x,t) \text{ if } x \in (\Omega \times \{0\}) \cup (\Omega \times \{l\}) \tag{29}$$

and the initial conditions

$$w(x,0) = T_0(x), \quad v(x,0) = u_0(x). \tag{30}$$

If the data of the problem (6)-(13) are smooth and the solutions of nonlinear problem (26)-(30) exists and is also smooth then we deduce from the relations (16) and (17) that the boundary conditions (8), (9) and the initial conditions (13) are satisfied for the asymptotic approximation $T^{(2)}$, $u^{(2)}$ up to $O(\varepsilon)$. Mention that $u^{(2)}$ has a form

$$u^{(2)}(x,t) = v(x,t) - \varepsilon \sum_{l=1}^{2} x_l \frac{\partial v}{\partial x_l}(x,t) + \varepsilon^2 u_2(x, \frac{\overline{x}}{\varepsilon}, t).$$

Thus we have constructed the formal asymptotic solution $T^{(2)}$, $u^{(2)}$ of the problem (6)-(13). The main term of it is a solution of the so called homogenized problem (26)-(30), which describes the macroscopic behavior of the processes in th e catalytic converter. It is much simpler than the initial problem (6)-(13): it does not depend on the rapid variables; the equation (27) is monodimensional with respect to space variables; the homogenized problem is of a parabolic type. Therefore th e homogenized model can be easily calculated by means of numerical analysis. All coefficients of the homogenized problem are explicitly related to the coefficients of the initial problem with the exception of $\hat{a}_{lq}$: these coefficients depe nd on the solution of so called cell problem ([2]) (23). The numerical solution of the cell problem is the main difficulty of the homogenization approach, but nowadays there exist an effective solver EFMODUL which calculates the effective (! macroscopic) characteristics of non-homogeneous structures. It was developed by one of the authors of the present article. It allows to complete the construction of the homogenized model of the catalytic converter.

References

[1] J.P. LECLERC, D. SCHWEICH, Modeling catalytic monoliths for automobile emission control, in Chemical Reactor Technology for Environmentally Safe Reactors and Products, Lara and all (eds), Kluwer , 1993, 547 - 576 .

[2] N.S. BAKHVALOV, G.P.PANASENKO, Homogenization : Averaging processes in periodic media, Kluwer , Dordrecht/Boston/London, 1989.

[3] G.P.PANASENKO, Numerical solution of cell problems in averaging theory, Zh.Vyc.Mat.i Mat.Fis.(ZVMMF), 1988, 28, 281-286.(in Russian) English transl. USSR Comput.Maths. Math.Phys., 1988, 28, No 1, 183-186.

CARLOS CONCA and JUAN DAVILA

Optimal bounds for mixtures of infinitely many materials

Abstract: In this paper we study optimal bounds for mixtures of infinitely many materials. Our physical motivation arises from optimal design of plates. The deflection of a symmetric plate with variable thickness h can be described by the classical Kirchhoff model, where the tensor of bending moments depends non linearly on the thickness h. Given a load on the plate, we try to minimize the work done by this force over a suitable set of thicknesses. Using homogenization techniques and a duality argument, this minimization is reduced to finding optimal bounds for the mixture of a family of plates. We study also the analogue of this optimal design problem for the diffusion equation, where we characterize partially the H-closure of a given set of materials and prove that the optimal composite can be appropiately described using a lamination procedure.

1   Introduction

Optimal design problems have usually no solution. One possible approach to overcome this difficulty consists in trying to find the so-called relaxed problem. In optimal design this can be performed by the homogenization method, the heart of it being the definition of H-convergence or G-convergence. Doing that, it appears naturally the problem of finding the closure with respect to H-convergence of a particular set of materials. This closure is called the set of generalized materials and it is generally very difficult to obtain an explicit characterization of it. But, when the functional to be minimized has a particular form, it has been possible to prove that an optimal generalized material can be described using a lamination procedure, even in situations (as linear elasticity) where the H-closure problem has not been solved yet. This is done by finding the optimal bounds for the generalized energy and the complementary energy quadratic forms.

In most of the previous works on this subject, attention was focused on mixtures of two, or at most a finite number of materials. Our purpose is to generalize these results to the case of mixing infinitely many materials.

Our motivation comes from the theory of plates and more exactly, it concerns the optimal design of a plate with variable thickness. We don't deal here with the physical aspects of the problem; for these we refer the reader to the work of Kohn and Vogelius [12]. We will just start providing some of the modelling ingredients of this problem and its precise mathematical formulation.

Let Ω be an open bounded subset of $I\!\!R^2$ with a smooth boundary and let us consider pure bending of a symmetric plate with midplane Ω and variable thickness $2h(x)$. The vertical displacement ω satisfies the classical Kirchhoff model, which is the following

59

fourth order equation

$$\partial_{\alpha\beta}(M_{\alpha\beta\gamma\delta}\partial_{\gamma\delta}\omega) = F \quad \text{in} \quad \Omega, \tag{1}$$

where $M_{\alpha\beta\gamma\delta}$ is given by

$$M_{\alpha\beta\gamma\delta}(x) = \frac{2}{3}h^3(x)B_{\alpha\beta\gamma\delta}.$$

As usual, we use here the standard convention of summation over repeated indices. Greek indices take the values 1 and 2, and we write $\partial_{\alpha\beta}$ as a shorthand of $\frac{\partial}{\partial x_\alpha}\frac{\partial}{\partial x_\beta}$.

The coefficients $B_{\alpha\beta\gamma\delta}$ depend on the material of which the plate is made, and assuming isotropy we have

$$\begin{aligned}
B_{1111} = B_{2222} &= \tfrac{E}{1-\nu^2} \\
B_{1122} = B_{2211} &= \tfrac{E\nu}{1-\nu^2} \\
B_{1212} = B_{1221} = B_{2112} = B_{2121} &= \tfrac{E}{2(1+\nu)}
\end{aligned} \tag{2}$$

where ν is the Poisson's ratio and E the Young's modulus of the material, and all other coefficients are zero.

The work done by the load F is given by $\int_\Omega F\omega \, dx$, and we will think of it as a function of h by fixing boundary conditions for ω, as for example

$$\omega = \frac{\partial \omega}{\partial n} = 0 \text{ on } \partial\Omega, \tag{3}$$

which corresponds to a completely campled plate.

It is physically interesting to minimize the work done among certain plates of prescribed volume, that is

$$\min \int_\Omega F\omega \, dx, \tag{4}$$

where ω is the solution of (1) with boundary conditions (3), and h is a measurable function satisfying $\int_\Omega h(x)dx = V_0$ and $h_{\min} \leq h(x) \leq h_{\max}$ a.e. in Ω. Here the constants are required to satisfy $0 < h_{\min} < h_{\max}$ and $h_{\min}\lceil\Omega| < V_0 < h_{\max}|\Omega|$.

Several authors have studied the case where the thickness h is a function of only one variable, because in this case it is possible to characterize the H-convergence of a sequence of tensors $h_\varepsilon^3 B$ in terms of weak-* convergence of some non linear functions of h_ε (see [5]). Using this result, Bonnetier and Conca [6, 7] have found a family of *total* relaxations of the optimal plate design problem.

In this paper we begin studying an optimal design problem which is the analogue of (4) in the case of the diffusion equation. Some of the reasons that make it interesting are that it generalizes the work of Murat and Tartar [15] to the case of mixing infinitely many materials and that the computations turn out to be much simpler than in the situation for plates.

Let B denote a symmetric, positive definite matrix, and Ω be an open bounded subset of $\mathbb{R}^N$. Consider a measurable function $h(x)$ satisfying $h_{\min} \leq h(x) \leq h_{\max}$ a.e in Ω. Let u be the unique solution of

$$\begin{cases} -\operatorname{div}(h^3(x)B\nabla u) &= f \quad \text{in} \quad \Omega \\ u &= 0 \quad \text{on} \quad \partial\Omega, \end{cases}$$

and let us define $L(h) = \int_\Omega fu$. In this simpler case, we are interested in minimizing $L(h)$ in the following set of admissible thicknesses:

$$\mathcal{H} = \left\{ h \in L^\infty(\Omega) \mid \int_\Omega h(x)dx = V_0 \text{ and } h_{\min} \leq h(x) \leq h_{\max} \text{ a.e in } \Omega \right\}.$$

The purpose of our work is to use homogenization for studying both of these optimal design problems. In the diffusion case we give a partial characterization of the H-closure of $\mathcal{H}$, compute optimal bounds and prove that, even for functionals of the type

$$L(h) = \int_\Omega g(x, u(x)) \, dx,$$

it is possible to find an optimal matrix with a laminate microstructure.

For the plate problem, attention is focused on the optimal bounds, which prove to be much harder to compute than in the difussion case. We find lower optimal bounds for the elastic and complementary energies of a generalized plate under some suitable hypothesis. The proof of optimality proceeds with the rank-n laminated materials introduced by Francfort and Murat [9]. Let us mention that optimal bounds have been obtained for mixtures of two incompressible, isotropic materials in [11] (see also [1]); for mixtures of two well-ordered isotropic materials in [2, 3] and for the analogous of plate theory in [10]. Our study pretends to generalize this latter paper to mixtures of infinitely many isotropic materials which depend non linearly on the thickness. It is worth mentioning that Avellaneda's works [2, 3] also discusses the optimal upper and lower bounds for a sum of elastic energies.

2 The homogenization method

It is widely recognized that optimal design problems like those mentioned in the introduction might have no solution and that a relaxation of them must be introduced. By a relaxation of

$$\min_{h \in \mathcal{H}} L(h)$$

we mean a set $\widetilde{\mathcal{H}}$ of the so called generalized admissible designs and a generalized functional $\widetilde{L}$ such that
  a) $\mathcal{H} \subseteq \widetilde{\mathcal{H}}$ and $\widetilde{L}|_{\mathcal{H}} = L$,

  b) for each $\widetilde{h} \in \widetilde{\mathcal{H}}$ there exists a sequence $(h_n)_n$ in $\mathcal{H}$ for which $\lim_{n \to \infty} L(h_n) = \widetilde{L}(\widetilde{h})$,

c) $\tilde{L}$ attains its minimum value on $\tilde{\mathcal{H}}$.

One way to get a relaxation is to use the classical techniques from homogenization theory (see the books of Benssousan, Lions and Papanicolaou [4] and Sánchez-Palencia [16]), and the proper tool is the definition of H-convergence introduced by F. Murat in 1978 [14]. Topologically speaking, by this definition two materials are close to each other if their responses to any external excitation are close to each other too.

DEFINITION 2.1. *(H-convergence) Let $\Omega \subseteq {I\!\!R}^N$ be an open set and $(A^\varepsilon)_\varepsilon$ a sequence in $L^\infty(\Omega)^{N^2}$ such that $A^\varepsilon(x)\xi \cdot \xi \geq \alpha |\xi|^2$ and $|A^\varepsilon(x)\xi| \leq \beta |\xi|$ a.e. in Ω (with $0 < \alpha < \beta$). We say in this case that the sequence $(A^\varepsilon)_\varepsilon$ is uniformly α-coercive and β-bounded. This sequence is said to H-converge to a coercive and bounded matrix A^* if the following statement holds true: for every ω open and bounded with closure contained in Ω and $f \in H^{-1}(\omega)$, the solution u^ε of*

$$\begin{cases} -\operatorname{div}(A^\varepsilon \nabla u^\varepsilon) &= f \quad in \quad \omega \\ u^\varepsilon &= 0 \quad on \quad \partial\omega, \end{cases}$$

satisfies $u^\varepsilon \rightharpoonup u^*$ in $H_0^1(\omega)$-weakly and $A^\varepsilon \nabla u^\varepsilon \rightharpoonup A^* \nabla u^*$ in $L^2(\omega)^N$-weakly, where u^* is the solution of

$$\begin{cases} -\operatorname{div}(A^* \nabla u^*) &= f \quad in \quad \omega \\ u^* &= 0 \quad on \quad \partial\omega. \end{cases}$$

The first step in using homogenization is to compute the H-closure of the set $\{h^3 B \mid h \in \mathcal{H}\}$, that is, to find all possible H-limits of sequences of matrices $A^\varepsilon = h_\varepsilon^3 B$, where $h_\varepsilon \in \mathcal{H}$. It is extremely difficult to get an explicit characterization of this closure, but we prove a theorem that sheds light on its structure. This result, the proof of which is given in [8], is analogous to Proposition 10 in the work of Murat and Tartar [15], but it is not so explicit.

THEOREM 2.1. *There exist compact sets S_h, $h \in [h_{\min}, h_{\max}]$, which do not depend on Ω, such that if A^* is the H-limit of a sequence $A^\varepsilon = h_\varepsilon^3 B$ with $h_\varepsilon \in \mathcal{H}$, then*

$$A^*(x) \in S_{h^*(x)} \quad a.e. \ in \quad \Omega, \tag{5}$$

where h^* is the weak-$*$ limit of the sequence $(h_\varepsilon)_\varepsilon$. Conversely, for any measurable A^* that verifies (5) with $h^* \in \mathcal{H}$ it is possible to construct a sequence $(h_\varepsilon)_\varepsilon$ contained in $\mathcal{H}$ such that $A^\varepsilon = h_\varepsilon^3 B$ H-converges to A^* and h_ε converges to h^* weakly-$*$.

If B is the identity matrix, then S_h contains only symmetric matrices, and for A, B symmetric with the same eigenvalues, $A \in S_h$ iff $B \in S_h$.

Thanks to Theorem 2.1 and the following well known characterization of the compliance

$$\int_\Omega fu \, dx = \inf_{\substack{\tau: \\ -\operatorname{div}\tau = f}} \int_\Omega A^{-1}\tau \cdot \tau \, dx,$$

where u is the unique solution of

$$\begin{cases} -\operatorname{div}(A\nabla u) &= f \quad in \quad \Omega \\ u &= 0 \quad on \quad \partial\Omega, \end{cases}$$

it follows that a total relaxation for the diffusion equation problem is given by

$$\min_{\substack{\tau: \\ -\mathrm{div}\tau = f}} \min_{h \in \mathcal{H}} \min_{A} \int_{\Omega} A^{-1}\tau \cdot \tau \; dx. \tag{6}$$

The last minimization in (6) ranges over all measurable functions A for which $A(x) \in S_{h(x)}$ a.e. in Ω. Remark that this can be done pointwise, i.e. it is enough to solve

$$\min_{A \in S_{\bar{h}}} A^{-1}\xi \cdot \xi \tag{7}$$

for any $\bar{h} \in [h_{\min}, h_{\max}]$ and $\xi \in \mathbb{R}^N$. This is equivalent to maximize $A\xi \cdot \xi$ on $S_{\bar{h}}$, or in other words, to compute the optimal upper bound for the elastic energy. For any $\xi \in \mathbb{R}^N$ and $A^* \in S_{\bar{h}}$, the energy $A^*\xi \cdot \xi$ can be bounded by the following quantities PROPOSITION 2.1.

$$\frac{1}{\theta(\bar{h})h_{\min}^{-3} + (1 - \theta(\bar{h}))h_{\max}^{-3}} \; B\xi \cdot \xi \;\; \le \;\; A^*\xi \cdot \xi \tag{8}$$

$$A^*\xi \cdot \xi \;\; \le \;\; \left(\theta(\bar{h})h_{\min}^3 + (1 - \theta(\bar{h}))h_{\max}^3\right) B\xi \cdot \xi \tag{9}$$

where $\theta(\bar{h}) = (h_{\max} - \bar{h})/(h_{\max} - h_{\min})$.
It is also possible to prove that these bounds are optimal in the sense that given any vector ξ and $\bar{h} \in [h_{\min}, h_{\max}]$ there are microstructures achieving equality in (8) and (9). DEFINITION 2.2. *A matrix A^* is said to have a laminate microstructure if it is the (constant) H-limit of a sequence*

$$A^\varepsilon(x) = \chi(x \cdot k/\varepsilon)h_1^3 B + (1 - \chi(x \cdot k/\varepsilon))h_2^3 B$$

with $h_1, h_2 \in [h_{\min}, h_{\max}]$ and $\chi : \mathbb{R} \to \{0, 1\}$ a periodic function. $k \in \mathbb{R}^N \setminus \{0\}$ is called the direction of lamination.
PROPOSITION 2.2. *For any $\xi \in \mathbb{R}^N$ and $\bar{h} \in [h_{\min}, h_{\max}]$ there are laminate microstructures in $S_{\bar{h}}$ achieving equality in (8) and (9). Moreover for these microstructures $h_1 = h_{\min}$ and $h_2 = h_{\max}$.*
It follows that $\min_{A \in S_{\bar{h}}} A^{-1}\xi \cdot \xi$ can be computed explicitly and moreover, there exists a matrix having a laminated microstructure that attains the minimum. This matrix can be described by a non zero vector, normal to the layers, and by a parameter h which can be interpreted as the mean value of the layers. To summarize THEOREM 2.2. *The relaxed problem (6) has a solution (τ^*, h^*, A^*) such that at any point $x \in \Omega$ $A^*(x)$ has a laminate microstructure.*

3 Sufficiency of extremal laminated materials

In the situation studied by Murat and Tartar [15] the admissible conductive tensors are allowed to take only two values. In our case these have the form $h^3 B$ with $h \in \mathcal{H}$, so that they can take a continuum of values. We could restrict h to be finite valued, and it is interesting to know whether our choice makes the functional reach a lower infimum or not. We work here with a more general functional than in the previous sections by letting L be

$$L(h) = \int_\Omega g(x, u(x))dx,$$

where $h \in \mathcal{H}$ and u is the unique solution of

$$\begin{cases} -\operatorname{div}(h^3(x)B\nabla u(x)) &= f \quad \text{in} \quad \Omega \\ u &= 0 \quad \text{on} \quad \partial\Omega. \end{cases}$$

The function $g(x, u)$ is required to satisfy the usual Carathéodory condition, i.e., measurability with respect to x and continuity in u, and the estimate

$$|g(x, u)| \le k(x) + c\,|u|^2\,,$$

with $k \in L^1(\Omega)$. These hypotheses make the mapping $u \to g(x, u(x))$ well defined from $H_0^1(\Omega)$ to $L^1(\Omega)$, and moreover, weak-strong sequentially continuous. A relaxation of

$$\min_{h \in \mathcal{H}} L(h)$$

is given by $\widetilde{\mathcal{H}} = \overline{\{(h, h^3 B) \mid h \in \mathcal{H}\}}$ and $\widetilde{L}(h, A) = \int_\Omega g(x, u(x))dx$, where $u \in H_0^1(\Omega)$ satisfies $-\operatorname{div}A\nabla u = f$ in Ω. THEOREM 3.1. *There is an optimal pair (h, A) for the relaxed problem such that, at every point x, $A(x)$ has a laminate microstructure, the layers being made of $h_{\min}^3 B$ and $h_{\max}^3 B$.*
The idea of the proof, following [15], is to choose any optimal solution (h^*, A^*) of the relaxed problem, and to replace it by (h, A) such that $A(x)$ has a laminate microstructure a.e. in Ω and $A^*\nabla u^* = A\nabla u^*$ a.e., where u^* is the unique solution in $H_0^1(\Omega)$ of $-\operatorname{div}A^*\nabla u^* = f$ in Ω.

The key result is the following lemma, which is the analogue of the lemma after Remark 32 in [15]. (See its proof in [8]). LEMMA 3.1. *Let e, e' be two vectors in $\mathbb{R}^N$ and let us define*

$$I(e, e') = \{h \in [h_{\min}, h_{\max}] \mid \exists A \in S_h \text{ such that } Ae = e'\}$$

Then, on the set where $I(e, e') \neq \emptyset$, the functions $\max I(e, e')$ and $\min I(e, e')$ are well defined and measurable, and for $h = \max I(e, e')$ or $h = \min I(e, e')$ there is a matrix $A \in S_h$ which is built using laminations between $h_{\min}^3 B$ and $h_{\max}^3 B$ such that $Ae = e'$.

4 Lower bounds for generalized plates

This section is devoted to compute lower bounds on the energy of the generalized plates and to investigate their optimality. Such bounds have been computed in several situations, as in elasticity, for a mixture of two components, or for the diffusion

64

equation (see Section 1 for references). In the latter case, an explicit characterization of the H-closure of the mixture of two conductors has been given (see [15]). Describing the H-closure remains an open question for elasticity.

Our situation concerning plates differs in two aspects. First, we are considering a fourth-order operator for a scalar field (the deflection). The differential relations that constrain admissible strain fields and fields of bending moments, turn out to be simpler to handle in the corresponding problem in elasticity. Secondly, we let the thickness take a continuum of values, instead of mixing only two materials. Moreover, the volume (corresponding to the proportion in the context of diffusion or elasticity) is a linear function of the thickness, whereas the bending tensor involves h^3.

4.1   The lower bound via the translation method

This bound is derived using the well known translation method of Milton [13], following the presentation of [1]. To this effect, let Y be an open cube in $\mathbb{R}^N$ and $\mathcal{M}_s$ the space of N by N symmetric matrices. Let M^* be an effective tensor obtained as the H-limit of a periodic sequence

$$M_\varepsilon = h_\varepsilon^3(x)B \qquad h_\varepsilon(x) = h(x/\varepsilon)$$

where $h : \mathbb{R}^N \to [h_{\min}, h_{\max}]$ is Y-periodic. The translation bound is a generalization of the well-known harmonic mean bound

$$M^* \geq \left(\frac{1}{|Y|} \int_Y M^{-1} \right)^{-1} \quad \text{with } M(y) = h^3(y)B.$$

DEFINITION 4.1. *A constant tensor τ is called quasiconvex if*

$$\int_Y \tau D^2 \varphi : D^2 \varphi \geq 0 \quad \text{for all } Y\text{-periodic } \varphi \in H^2(Y).$$

REMARK 4.1. *Let τ be a constant fourth order tensor. Then τ is quasiconvex if and only if*

$$\tau k \otimes k : k \otimes k \geq 0 \quad \forall k \in \mathbb{R}^N. \tag{10}$$

PROPOSITION 4.1. *Let τ be a quasiconvex tensor that satisfies the following condition:*

$$M - \tau \geq 0 \text{ a.e. in } Y. \tag{11}$$

Then

$$M^* - \tau \geq \left(\frac{1}{|Y|} \int_Y (M - \tau)^{-1} \right)^{-1}.$$

We apply now last proposition by considering the special translation τ_η given by

$$\tau_\eta = M_{\min} - g(\eta)^{-1}\eta \otimes \eta,$$

where

$$M_{\min} = h_{\min}^3 B \qquad g(\eta) = \sup_{k \neq 0} \frac{(\eta : k \otimes k)^2}{M_{\min}k \otimes k : k \otimes k}.$$

It is straightforward to check that τ_η is quasiconvex and satisfies (11). Therefore, using Proposition 4.1, one gets the inequality

$$\frac{1}{1 + g(\eta)^{-1}(M^* - M_{\min})^{-1}\eta : \eta} \geq \frac{1}{|Y|} \int_Y \frac{1}{1 + g(\eta)^{-1}(h^3 - h_1^3)^{-1}B^{-1}\eta : \eta} \, dy \quad (12)$$

In order to obtain a bound independent of the average volume of the plate, we minimize the right hand side of (12) under the constraint

$$\frac{1}{|Y|} \int_Y h \, dy = \bar{h}.$$

This is done by finding the convex hull of the graph of the function

$$h \rightarrow \frac{1}{1 + g(\eta)^{-1}(h^3 - h_1^3)^{-1}B^{-1}\eta : \eta}.$$

To simplify this, we assume here concavity of this function, which holds true when

$$\inf_{k \neq 0} \frac{B^{-1}\eta : \eta Bk \otimes k : k \otimes k}{(\eta : k \otimes k)^2} \leq 3 \quad (13)$$

After elementary manipulations one arrives at

$$\theta(\bar{h})g(\eta) + (M_{\max} - M_{\min})^{-1}\eta : \eta \geq (1 - \theta(\bar{h}))(M^* - M_{\min})^{-1}\eta : \eta,$$

where $\theta(\bar{h})$ is $(h_{\max} - \bar{h})/(h_{\max} - h_{\min})$ and $M_{\max} = h_{\max}^3 B$. Observing now that

$$(1 - \theta(\bar{h}))^2(M^* - M_{\min})^{-1}\eta : \eta = \sup_{\xi \in \mathcal{M}_s} 2(1 - \theta(\bar{h}))\xi : \eta - (M^* - M_{\min})\xi : \xi,$$

one finally gets the estimate

$$(M^* - M_{\min})\xi : \xi \geq \sup_\eta (1 - \theta(\bar{h})) \left[2\xi : \eta - (M_{\max} - M_{\min})^{-1}\eta : \eta - \theta(\bar{h})g(\eta) \right] \quad (14)$$

REMARK 4.2. *It is interesting to know whether hypothesis (13) holds in some special cases. We notice that when B is the identity tensor, then*

$$\inf_{k \neq 0} \frac{B^{-1}\eta : \eta Bk \otimes k : k \otimes k}{(\eta : k \otimes k)^2} \leq N,$$

where N is the space dimension, so that in this case (13) is true for $N \leq 3$.

66

4.2 Optimality of the lower bound for the energy

In this section we find generalized plates achieving equality in the bound found in the previous section. These generalized plates are the analogues of the laminated composites considered in diffusion or elasticity. Here the term laminated plate may be misleading, since the layers are distributed in the plane xy and do not refer to a "sandwich structure" across the thickness of the plate.

Given two constant, symmetric tensors $M_1 = h_1^3 B$ and $M_2 = h_2^3 B$ and a direction $k \in S^1$, the H-limit of the sequence

$$M^\varepsilon = \chi(x \cdot k/\varepsilon) M_1 + (1 - \chi(x \cdot k/\varepsilon)) M_2, \qquad \chi \text{ 1-periodic}$$

is called a rank-1 stiffened plate, or simply a rank-1 laminate. It is characterized by h_1, h_2, the direction k, and the proportion

$$\theta = w * - \lim_{\varepsilon \to 0} \chi(\cdot/\varepsilon),$$

as the following proposition shows : PROPOSITION 4.2. *If $h_1 \neq h_2$, the effective tensor of the rank-1 stiffened plate in the direction k and with proportion θ is given by*

$$(1 - \theta)(M_1^* - M_1)^{-1} b = (M_2 - M_1)^{-1} b + \theta \frac{b : k \otimes k}{M_1 k \otimes k : k \otimes k} \qquad (15)$$

By induction we define a rank-n laminate ($n \geq 2$), with directions $k_1, ..., k_n$ and proportions $\rho_1, ..., \rho_n$ and reference materials M_1 and M_2 as the H-limit M_n^* of the sequence

$$M^\varepsilon(x) = \chi_n(x \cdot k_n/\varepsilon) M_1 + (1 - \chi_n(x \cdot k_n/\varepsilon)) M_{n-1}^*,$$

where M_{n-1}^* is a rank-$(n-1)$ laminate with directions $k_1, ..., k_{n-1}$ and proportions $\rho_1, ..., \rho_{n-1}$, χ_n is 1-periodic and $w * - \lim_{\varepsilon \to 0} \chi_n(\cdot/\varepsilon) = \rho_n$. As noted in [9], the effective tensor of a rank-n laminate is easily computed inductively from the expression (15) : PROPOSITION 4.3. *The effective tensor M_n^* is given by*

$$(1 - \theta)(M_n^* - M_1)^{-1} = (M_2 - M_1)^{-1} + \theta \sum_{i=1}^{n} m_i f_{M_1}(k_i),$$

where

$$f_{B_1}(k) b = \frac{b : k \otimes k}{B_1 k \otimes k : k \otimes k} k \otimes k \qquad \theta = 1 - \prod_{i=1}^{n} 1 - \rho_i$$

$$m_1 = \rho_1/\theta \qquad m_i = \left[\prod_{j=1}^{i-1} 1 - \rho_j\right] \rho_i/\theta \quad for \; i \geq 2.$$

THEOREM 4.1. *For all $\bar{h} \in [h_{\min}, h_{\max}]$ and for any $\xi \in \mathcal{M}_s$ there is a rank-p laminate plate M^* which attains the bound (14).*

The proof of this theorem is based on a standard argument of optimality in the following concave maximization (see e.g. [1]):

$$\sup_{\eta \in \mathcal{M}_s} 2\xi : \eta - (M_{\max} - M_{\min})^{-1} \eta : \eta - \theta(\bar{h}) g(\eta).$$

4.3 Lower bound for the complementary energy

For the plate design problem it is interesting to compute the lower bound for the complementary energy. THEOREM 4.2. *Consider a tensor B defined as in (2), where E is the Young's modulus and ν the Poisson's ratio. Then there exists $\varepsilon > 0$ such that for $|\nu| \le \varepsilon$ and $h_{\min}/h_{\max} \le \varepsilon$, the following estimate holds*

$$(M^{*-1} - M_{\max}^{-1})\xi : \xi \ge$$

$$\ge \sup_{\eta} \theta(\bar{h}) \left[2\xi : \eta - (M_{\min}^{-1} - M_{\max}^{-1})^{-1}\eta : \eta - (1 - \theta(\bar{h}))g_c(\eta) \right]$$

for all $\xi \in \mathcal{M}_s$.

Acknowledgements. Both authors gratefully acknowledge the support of the Chilean Programme of Presidential Chairs in Sciences and Conicyt through Fondecyt Grant № 1970734.

References

[1] ALLAIRE G., KOHN R. V., *Optimal bounds on the effective behavior of a mixture of two well-ordered elastic materials*, Quart. Appl. Math. **51**, pp. 643-679 (1993).

[2] AVELLANEDA M., *Optimal bounds and microgeometries for elastic two-phase composites*, SIAM J. Appl. Math. **47**, pp. 1216-1228 (1987).

[3] AVELLANEDA M., *Bounds on the effective elastic constants of two-phase composites materials*, in "Nonlinear Partial Differential Equations and their Applications", Collège de France Seminar vol. X, H. Brezis and J. L. Lions eds., Pitman Research Notes in Mathematics Series, Longman, Harlow (1989).

[4] BENSOUSSAN A., LIONS J.L., PAPANICOLAOU G., *Asymptotic Analysis for Periodic Structures*, North Holland, Amsterdam (1978).

[5] BONNETIER E., VOGELIUS M., *Relaxation of a compliance functional for a plate optimization problem*, in "Applications of Multiple Scaling in Mechanics", P. G. Ciarlet and E. Sánchez-Palencia eds., Masson, Paris (1987).

[6] BONNETIER E., CONCA C., *Approximation of Young measures by functions and application to a problem of optimal design for plates with variable thickness*, Proc. Roy. Soc. Edinburgh Sect. A **124**, pp. 399-422 (1994).

[7] BONNETIER E., CONCA C., *Relaxation totale d'un problème d'optimisation de plaques*, C. R. Acad. Sci. Paris Sér. I Math. t **317**, pp. 931-93 (1993).

[8] BONNETIER E., CONCA C., DÁVILA J., *Homogenization and optimal bounds for mixtures of an infinite family of materials* (working paper).

[9] FRANCFORT G., MURAT F., *Homogenization and optimal bounds in linear elasticity*, Arch. Rational Mech. Anal. **94**, pp. 307-334 (1986).

[10] GIBIANSKI L., CHERKAEV A., *Design of composites plates of extremal rigidity*, Ioffe Physicotechnical Institute, preprint (1984). (English translation to appear in "Topics in the Mathematical Modelling of Composite Materials", R.V. Kohn eds., series "Progress in Nonlinear Differential Equations and their Applications", Birkhauser, Boston).

[11] KOHN R.V., LIPTON R., *Optimal bounds for the effective energy of mixtures of isotropic incompressible, elastic materials*, Arch. Rational Mech. Anal. **102**, pp. 331-350 (1988).

[12] KOHN R.V., VOGELIUS M., *Thin plates with rapidly varying thickness, and their relation to structural optimization*, in "Homogenization and Effective Moduli of Materials and Media", J. Ericksen & al. eds., pp. 126-149, Springer Verlag (1986).

[13] G. MILTON, *On characterizing the set of possible effective tensors of composites: the variational method and the translation method*, Comm. Pure Appl. Math. **43**, pp. 63-125 (1990).

[14] MURAT F., TARTAR L., *H-convergence*, Séminaire d' Analyse Fonctionnelle et Numérique de l' Université d' Alger, mimeographed notes (1977-78) (English translation to appear in "Topics in the Mathematical Modelling of Composite Materials",R. V. Kohn eds., series "Progress in Nonlinear Differential Equations and their Applications", Birkhauser, Boston).

[15] MURAT F., TARTAR L., *Calcul des variations et homogénéisation*, in "Méthodes de l'Homogénéisation: Théorie et Applications en Physique", Eyrolles, pp. 319-369 (1985).

[16] SÁNCHEZ-PALENCIA E., *Nonhomogeneous media and vibrating theory.* Lecture Notes in Physics **127**, Springer-Verlag, Berlin, (1980).

ANAHÍ DELLO RUSSO* and RODOLFO RODRÍGUEZ[†]

Finite element displacement methods for fluid-solid vibration problems

1 Introduction

Increasing attention has been recently paid to the numerical computation of fluid-solid interactions (see, for instance, [7] and the references therein). The present paper deals with the so-called *interior elastoacoustics problem* which concerns the determination of the small amplitude motions of a bounded inviscid compressible fluid interacting with an elastic solid; in particular, the computation of the free vibration modes of such coupled system.

Finite element methods are typically used for this problem. The equations for the solid are generally expressed in terms of displacements, whereas different variables have been used for the fluid: pressure, potential or displacements. The drawback of the two former is that, because of the interface conditions, after discretization they lead to nonsymmetric eigenvalue problems. Instead, the use of displacement variables for both, the fluid and the solid, yields a sparse symmetric eigenvalue problem.

However, it is well known that the use of a displacement formulation in the fluid combined with classical Lagrangian finite elements leads to the presence of spurious or circulation modes; i.e., non-irrotational modes which have no physical entity. These spurious modes may correspond either to zero or to non-zero frequencies and it is very difficult to separate them from those of the real physical problem.

A first attempt to circumvent this problem was made in [6]. It consists of penalizing the curl-free condition. This technique does not eliminate the nonphysical modes but they are pushed towards higher frequencies and, therefore, do not appear among the first ones (see [4] for a mathematical proof).

More recently, Chen and Taylor[5] proposed to use bilinear rectangular elements for the fluid displacements combined with a reduced integration in the stiffness matrix of the fluid and a projection of the corresponding element mass matrices. They exhibit numerical experiments showing that this procedure is useful to eliminate the spurious modes, but no proof of the effectiveness of this method is known to date.

Another approach was considered in [4]. It consists of using lowest order Raviart-Thomas triangular elements to discretize the fluid displacements, the coupling at the fluid-solid interface being taken into account in a weak sense. Mathematical aspects concerning convergence of the scheme, error estimates and the proof that spurious modes do not arise can be found in [1].

In this paper we first extend the analysis in [1, 4] to quadrilateral Raviart-Thomas

---
*Comisión de Investigaciones Científicas de la Provincia de Buenos Aires, Argentina.
[†]Partially funded by FONDECYT (Chile) under grant No. 1.969.615.

elements. By doing so, we are able to compare the performance of this method with that of Chen-Taylor's one on the same meshes. We have used as a test problem the computation of the free vibration modes of a compressible fluid contained in a solid, the walls of the solid being either perfectly rigid or membranes allowing for fluid-solid interaction. We show that both methods give quite similar results in the case of plane walls.

Then we consider solids with curved walls. We show that, in this case, the direct application of the kinematic conditions in Chen-Taylor's method would produce a spurious boundary layer for the tangential component of the fluid displacements near the curved boundaries, which would deteriorate the order of convergence. To avoid this drawback, we introduce an alternative way of imposing the kinematic restrictions in a weak sense. We show that such procedure allows for non-vanishing tangential displacements of the fluid along the curved walls. We present numerical experiments showing the efectiveness of this approach.

2 The model problem

We consider an inviscid compressible fluid contained in a vessel whose walls are either perfectly rigid or membranes. We neglect gravity effects as well as any other exterior forces. We denote by Ω the bounded domain occupied by the fluid, by $\partial\Omega$ the boundary of Ω and by n its outward unit normal vector. The surface $\partial\Omega$ is splitted into two parts, Γ_M and Γ_R; Γ_R is assumed to be rigid whereas Γ_M is an ideal membrane with its boundary $\partial\Gamma_M$ fixed to the rigid walls.

When displacement variables are used to describe the fluid, the arguments in [4] can be easily adapted to show that the small amplitude free vibration modes of the fluid-membrane coupled system are solutions of the following spectral problem:

Find a real number ω and functions $U \in H(\text{div}, \Omega)$ and $W \in H^1(\Gamma_M)$, not both identically zero, satisfying

$$
\begin{align}
-\nabla(\rho_F c^2 \nabla \cdot U) &= \omega^2 \rho_F U &&\text{in } \Omega &&(1)\\
-T_0 \nabla^2 W - \rho_F c^2 \nabla \cdot U &= \omega^2 \rho_M W &&\text{on } \Gamma_M &&(2)\\
W &= 0 &&\text{on } \partial\Gamma_M &&(3)\\
U \cdot n &= W &&\text{on } \Gamma_M &&(4)\\
U \cdot n &= 0 &&\text{on } \Gamma_R &&(5)
\end{align}
$$

In the previous equations ω is the vibration frequency, U and W are the amplitudes of the fluid displacement field and the transverse displacement of the membrane, respectively, ρ_F and ρ_M the respective densities, c is the sound speed of the fluid and T_0 the tension of the membrane at equilibrium.

In order to obtain a variational formulation of this problem we consider the space of admissible displacements

$$\mathcal{V} := \left\{ (U, W) \in H(\mathrm{div}, \Omega) \times H_0^1(\Gamma_{\mathrm{M}}) \; : \; U \cdot n = 0 \text{ on } \Gamma_{\mathrm{R}} \text{ and } U \cdot n = W \text{ on } \Gamma_{\mathrm{M}} \right\}.$$

Given a pair of admissible displacements $(Y, Z) \in \mathcal{V}$, by integrating equations (1) and (2) multiplied by Y and Z, respectively, we obtain the following variational formulation of the spectral problem (1-5) where we denote $\lambda = \omega^2$:

Find $\lambda \in \mathbb{R}$ *and* $(U, W) \in \mathcal{V}$, $(U, W) \neq 0$, *satisfying*

$$\int_\Omega \rho_{\mathrm{F}} c^2 \, \nabla \cdot U \, \nabla \cdot Y \, dx + \int_{\Gamma_{\mathrm{M}}} T_0 \, \nabla W \cdot \nabla Z \, d\Gamma$$

$$= \lambda \left(\int_\Omega \rho_{\mathrm{F}} U \cdot Y \, dx + \int_{\Gamma_{\mathrm{M}}} \rho_{\mathrm{M}} W Z \, d\Gamma \right), \qquad \forall (Y, Z) \in \mathcal{V}. \quad (6)$$

Notice that $\lambda = 0$ is an eigenvalue of this problem with associated eigenspace

$$\mathcal{K} := \{ (U, 0) \in \mathcal{V} \; : \; \mathrm{div}\, U = 0 \text{ in } \Omega \text{ and } U \cdot n = 0 \text{ on } \partial\Omega \}.$$

This is due to the fact that no irrotationality condition was assumed for the fluid in the spectral problem (1-5). The fluid displacements in $\mathcal{K}$ correspond to pure rotational motions inducing neither changes of pressure nor vibrations in the membrane. They are spurious solutions of this pure displacement formulation of the problem and they are the reason why spurious modes arise when standard linear finite elements are used to discretize the fluid.

Apart from $\lambda = 0$, the solutions of problem (6) are a sequence of positive eigenfrequencies λ_n of finite multiplicity with associated eigenfunctions (U_n, W_n) corresponding to irrotational fluid displacements. This can be easily proved by adapting to this case the arguments given in [1] for a vibration problem with fluid-elastic structure interaction.

3 Finite element discretizations

Two alternative discretizations have been recently introduced to avoid spurious modes in displacement formulations of fluid-solid vibration problems. Both have been described and numerically tested for 2D problems, but their extensions to 3D are straightforward.

The first one, by Chen and Taylor, consists of using bilinear quadrilateral elements combined with a reduced integration in the stiffness matrix of the fluid and a projection of the element mass matrix. Numerical experiments showing that this procedure is useful to eliminate the spurious modes are described in [5].

The second method[4] consists of using Raviart-Thomas triangular elements to discretize the fluid equation. The degrees of freedom of these elements are located at the edges of the triangles and represent the normal components of the field along them. Error estimates and the proof that spurious modes do not arise are given in [1].

No direct extension of Chen-Taylor's method to linear triangular elements seems to be possible. Instead, Raviart-Thomas elements for quadrilaterals are well known.[8] In order to compare the effectiveness of both methods, we first extend the analysis in [4] to this type of elements for the fluid-membrane vibration problem (6).

We describe the method for 2D fluid domains but the extension to 3D is also straightforward. Notice that, in this 2D case, the "membrane" reduces to a string. We consider a mesh of regular quadrilaterals for the fluid domain Ω; we denote by h the meshsize. For the membrane domain we consider the mesh given by those edges of quadrilaterals contained in Γ_M.

3.1 Discretization by Raviart-Thomas elements

In this case, the fluid displacement field is discretized by means of lowest order Raviart-Thomas elements. When defined in the reference square $\hat{K} := [-1, 1] \times [-1, 1]$ they are of the form $\hat{U}(\hat{x}, \hat{y}) = (a + b\hat{x}, c + d\hat{y})$. Such fields have constant normal component on each edge of the square $\hat{K}$. These normal components are the degrees of freedom used to determine the real numbers a, b, c, d defining $\hat{U}$.

Now, to define U_h on any quadrilateral K of the mesh, we use the Piola transform in order to preserve normal components. More precisely, let

$$
\begin{aligned}
F_K : \quad \hat{K} &\longrightarrow K \\
(\hat{x}, \hat{y}) &\mapsto (x, y)
\end{aligned}
\tag{7}
$$

be a bilinear map applying the four vertices of $\hat{K}$ on those of K (preserving its order) and, hence, the whole $\hat{K}$ onto K; for $(x, y) \in K$ we define

$$
U_h(x, y) = \frac{1}{\det J_K} J_K \hat{U}(\hat{x}, \hat{y}),
$$

where J_K is the Jacobian matrix of F_K. The field U_h, patchwise defined in this way, satisfies the following properties:[8]

$$
\nabla \cdot U_h(x, y) = \frac{1}{\det J_K} \hat{\nabla} \cdot \hat{U}(\hat{x}, \hat{y}),
$$

and

$$
\int_\ell U_h(x, y) \cdot n \, d\Gamma = \int_{\hat{\ell}} \hat{U}(\hat{x}, \hat{y}) \cdot \hat{n} \, d\hat{\Gamma}, \qquad \text{for each edge } \ell \text{ of } \partial K.
$$

In general, U_h have discontinuous tangential components along each edge of the quadrilaterals. However, their normal components across these edges are continuous and this is enough to ensure that $\nabla \cdot U_h$ is globally well defined.

The transverse displacement of the membrane is discretized by means of standard piecewise linear continuous elements W_h vanishing at $\partial \Gamma_M$. The boundary condition

74

(5) is directly imposed to the Raviart-Thomas' elements; however, this cannot be done with the interface condition (4) since, if only discrete displacements satisfying (4) were used, then the membrane would not be allowed to move. In fact, for each edge $\ell \subset \Gamma_M$, $U_h \cdot n|_\ell$ is constant. Hence, if $U_h \cdot n|_\ell = W_h|_\ell$ were assumed, $W_h|_\ell$ should be constant too. Therefore, since $W_h = 0$ on $\partial\Gamma_M$, because of the continuity of W_h, no membrane displacement would be possible.

Therefore, we are led to use a nonconforming approximation: we impose (4) in the following weak sense:

$$\int_\ell (U_h \cdot n - W_h)\, d\Gamma = 0, \qquad \text{for each edge } \ell \subset \Gamma_M. \tag{8}$$

Let us remark that, since $U_h \cdot n$ is constant and W_h is linear, (8) is equivalent to assume that $U_h \cdot n$ and W_h coincide at the midpoint of each edge $\ell \subset \Gamma_M$.

In order to describe an efficient way of implementing the linear constraint (8), it is convenient to consider an alternative variational formulation of the eigenvalue problem (1-5) involving the pressure of the fluid $P = -\rho_F c^2 \nabla \cdot U$. Let

$$\mathcal{X} := \left\{ (U, W) \in H(\text{div}, \Omega) \times H_0^1(\Gamma_M) : U \cdot n = 0 \text{ on } \Gamma_R \right\}$$

be the space of displacements not (neccesarily) satisfying the kinematic constraint on the interface. By integrating equations (1) and (2) conveniently multiplied by test functions in $\mathcal{X}$ and by imposing the interface condition (4) in a weak sense, the following hybrid variational formulation of the spectral problem (1-5) is obtained:

Find $\lambda \in \mathbb{R}$, and $(U, W, P) \in \mathcal{X} \times L^2(\Gamma_M)$, $(U, W, P) \neq 0$, satisfying

$$\int_\Omega \rho_F c^2 \nabla \cdot U \, \nabla \cdot Y \, dx + \int_{\Gamma_M} T_0 \nabla W \cdot \nabla Z \, d\Gamma + \int_{\Gamma_M} P\,(Y \cdot n - Z)\, d\Gamma$$

$$= \lambda \left(\int_\Omega \rho_F U \cdot Y \, dx + \int_{\Gamma_M} \rho_M W Z \, d\Gamma \right), \qquad \forall (Y, Z) \in \mathcal{X},$$

$$\int_{\Gamma_M} (U \cdot n - W)\, Q \, d\Gamma = 0, \qquad \forall Q \in L^2(\Gamma_M).$$

The function $P \in L^2(\Gamma_M)$ plays the role of a Lagrange multiplier for the kinematic constraint and its physical meaning is the pressure between the fluid and the membrane.

Now, we discretize the fluid displacement field by Raviart-Thomas' elements, the membrane displacements by piecewise linear elements and the interface pressure P by piecewise constant functions. Hence, we obtain the following discrete eigenvalue problem:

Find $\lambda_h \in \mathbb{R}$ and $(U_h, W_h, P_h) \in \mathcal{X}_h \times \mathcal{C}_h$, $(U_h, W_h, P_h) \neq 0$, satisfying

$$\int_\Omega \rho_F c^2 \nabla \cdot U_h \, \nabla \cdot Y_h \, dx + \int_{\Gamma_M} T_0 \nabla W_h \cdot \nabla Z_h \, d\Gamma + \int_{\Gamma_M} P_h\,(Y_h \cdot n - Z_h)\, d\Gamma$$

$$= \lambda_h \left(\int_\Omega \rho_{\scriptscriptstyle\mathrm{F}} U_h \cdot Y_h \, dx + \int_{\Gamma_{\scriptscriptstyle\mathrm{M}}} \rho_{\scriptscriptstyle\mathrm{M}} W_h Z_h \, d\Gamma \right), \quad \forall (Y_h, Z_h) \in \mathcal{X}_h, \qquad (9)$$

$$\int_{\Gamma_{\scriptscriptstyle\mathrm{M}}} (U_h \cdot n - W_h) \, Q_h \, d\Gamma = 0, \quad \forall Q_h \in \mathcal{C}_h, \qquad (10)$$

where $\mathcal{C}_h$ and $\mathcal{X}_h$ are the discrete spaces described above.

Any solution of problem (9-10) is a solution of the discrete problem associated with (6) with the kinematic constraint imposed as in (8); in fact, (10) ensures that the discrete displacements satisfy (8). Conversely, following the techniques in [2], it can be proved that the mixed problem (9-10) satisfy both of Brezzi's classical conditions and, hence, for any solution of the discrete problem associated with (6), there exists a piecewise constant function P_h such that (9) is satisfied.

Clearly, $\lambda_h = 0$ is an eigenvalue of problem (9-10) with corresponding eigenspace

$$\mathcal{K}_h := \{ (U_h, 0, 0) \in \mathcal{X}_h \times \mathcal{C}_h \; : \; \operatorname{div} U_h = 0 \text{ in } \Omega \text{ and } U_h \cdot n = 0 \text{ on } \partial\Omega \}.$$

It can be readily shown that these discrete divergence-free displacements are exactly those given by

$$U_h = \operatorname{curl} \varphi_h,$$

with φ_h being any isoparametric bilinear function defined in Ω and vanishing on $\partial\Omega$.

The techniques in [1] [2] and [9] can be adapted to problem (9-10) in order to prove that its eigenfrequencies converge to those of problem (6) and that non-zero frequency spurious modes do not arise in this discretization. More precisely we have the following result:

THEOREM 1. *Let $\Omega \subset \mathbb{R}^2$ be a polygonal domain. Let $r := 1$ if Ω is convex, or $r := \frac{2\pi}{\theta}$, with θ being the maximum nonconvex angle of $\partial\Omega$, otherwise.*

Let $\lambda_1 \leq \lambda_2 \leq \ldots \leq \lambda_n \leq \ldots$ and $\lambda_{h1} \leq \lambda_{h2} \leq \ldots \leq \lambda_{hN_h}$ be the strictly positive eigenvalues of problems (6) and (9-10), respectively (in both cases repeated according to their multiplicities).

1. The following error estimate holds:

$$|\lambda_n - \lambda_{hn}| \leq C h^{2r}. \qquad (11)$$

2. Let λ_n be a simple eigenvalue. Let (U_n, W_n) be a normalized associated eigenvector. Let (U_{hn}, W_{hn}) be a discrete eigenvector corresponding to the eigenvalue λ_{hn} normalized in the same manner. Then the following error estimate holds:

$$\|U_n - U_{hn}\|_{H(\mathrm{div},\Omega)} + \|W_n - W_{hn}\|_{H^1(\Gamma_{\scriptscriptstyle\mathrm{M}})} \leq C h^r. \qquad (12)$$

Notice that property (11) implies, in particular, that there are no spurious eigenvalues associated with non-zero frequencies. On the other hand, error estimates similar to (12) can be proved for multiple eigenvalues too. See [1, 9] for a precise statement.

76

Finally, we give a matricial description of problem (9-10) to show that it is a well posed symmetric generalized eigenvalue problem. Let us call $\tilde{U}_h$, $\widetilde{W}_h$ and $\tilde{P}_h$ the vectors of nodal components of the discrete eigenfunctions U_h, W_h and P_h, respectively. Problem (9-10) can be written in the following way:

$$
\begin{pmatrix} K_{\text{F}} & 0 & C \\ 0 & K_{\text{M}} & D \\ C^t & D^t & 0 \end{pmatrix} \begin{pmatrix} \tilde{U}_h \\ \widetilde{W}_h \\ \tilde{P}_h \end{pmatrix} = \lambda_h \begin{pmatrix} M_{\text{F}} & 0 & 0 \\ 0 & M_{\text{M}} & 0 \\ 0 & 0 & 0 \end{pmatrix} \begin{pmatrix} \tilde{U}_h \\ \widetilde{W}_h \\ \tilde{P}_h \end{pmatrix}, \tag{13}
$$

where K_{F} and M_{F}, K_{M} and M_{M}, are the stiffness and mass matrices of the fluid and the membrane, respectively, and C and D are the coupling matrices of the pressure at the interface with the fluid and the membrane, respectively; that is:

$$
\tilde{U}_h^t K_{\text{F}} \tilde{Y}_h = \int_\Omega \rho_{\text{F}} c^2 \nabla \cdot U_h \nabla \cdot Y_h \, dx, \qquad \widetilde{W}_h^t K_{\text{M}} \tilde{Z}_h = \int_{\Gamma_{\text{M}}} T_0 \nabla W_h \cdot \nabla Z_h \, d\Gamma,
$$

$$
\tilde{U}_h^t M_{\text{F}} \tilde{Y}_h = \int_\Omega \rho_{\text{F}} U_h \cdot Y_h \, dx, \qquad \widetilde{W}_h^t M_{\text{M}} \tilde{Z}_h = \int_{\Gamma_{\text{M}}} \rho_{\text{M}} W_h Z_h \, d\Gamma,
$$

$$
\tilde{U}_h^t C \tilde{P}_h = \int_{\Gamma_{\text{M}}} P_h U_h \cdot n \, d\Gamma, \qquad \widetilde{W}_h^t C \tilde{P}_h = \int_{\Gamma_{\text{M}}} P_h W_h \, d\Gamma.
$$

Both matrices in (13) are singular; however, by performing a translation in the eigenvalues, this problem can be written in the following more convenient equivalent way:

$$
\begin{pmatrix} K_{\text{F}} + M_{\text{F}} & 0 & C \\ 0 & K_{\text{M}} + M_{\text{M}} & D \\ C^t & D^t & 0 \end{pmatrix} \begin{pmatrix} \tilde{U}_h \\ \widetilde{W}_h \\ \tilde{P}_h \end{pmatrix} = (1 + \lambda_h) \begin{pmatrix} M_{\text{F}} & 0 & 0 \\ 0 & M_{\text{M}} & 0 \\ 0 & 0 & 0 \end{pmatrix} \begin{pmatrix} \tilde{U}_h \\ \widetilde{W}_h \\ \tilde{P}_h \end{pmatrix}. \tag{14}
$$

Following the techniques in [3], it is easy to prove that the matrix in the left hand side of (14) is non singular. Consequently, it yields a well-posed generalized eigenvalue problem with both matrices being symmetric and highly sparse.

3.2 Chen-Taylor's discretization

In this case, both components of the fluid displacements are discretized by means of isoparametric bilinear elements. That is, for each quadrilateral K, let $(x, y) = F_K(\hat{x}, \hat{y})$ be as defined in (7). The components of $U_h = (U_{h1}, U_{h2})$ are of the form

$$
U_{hi}(x, y) = \hat{U}_i(\hat{x}, \hat{y}) = A_i + B_i \hat{x} + C_i \hat{y} + D_i \hat{x} \hat{y}, \qquad i = 1, 2. \tag{15}
$$

The degrees of freedom used to determine the values of A_i, B_i, C_i and D_i $(i = 1, 2)$ are the values of the corresponding component at the vertices of K.

The transverse displacements of the membrane are discretized by means of standard piecewise linear continuous elements W_h as above. For a polygonal fluid domain Ω, the kinematic constraint (4) can now be directly imposed, leading to a conforming method.

In fact, since the normal component $U_h \cdot n|_\ell$ is now linear for each edge $\ell \subset \Gamma_{\mathrm{M}}$, there is no problem in considering discrete displacements satisfying this constraint. We discuss below how to impose the boundary condition (5) in the case of a curved rigid boundary Γ_{R}. By the moment, we restrict our attention to the case considered in [5] of a polygonal fluid domain.

A direct use of this discretization would yield a discrete eigenvalue problem with a non singular stiffness matrix. Therefore, $\lambda_h = 0$ would not be an eigenvalue of this discrete problem and, hence, many spurious rotational eigenmodes would arise asociated with non-zero frequencies. To avoid this problem, Chen and Taylor proposed in [5] to under-integrate the stiffness matrix. They have used a one-point Gaussian quadrature rule to compute each element stiffness matrix. This procedure amounts to eliminate in (15) the terms of the form $D_i \hat{x} \hat{y}$ in trial and test functions for the computation of the element stiffness matrices. These terms represent the so-called *hourglass* modes.

On the other hand, each element mass matrix is projected onto the orthogonal complement of the subspace spanned by the hourglass modes. Once more, this procedure amounts to not to consider the terms of the form $D_i \hat{x} \hat{y}$ in (15) for the computation of the element mass matrices. The numerical experiments in [5] show that, by doing so, all the eigenvalues of the discrete problem are good approximations of those of the continuous one and with the same multiplicity.

4 Numerical Experiments

4.1 A fluid-membrane vibration problem

As a first problem we have computed the natural vibration modes of a coupled system consisting of a compressible fluid contained in a rectangular 2D cavity with its upper wall being an ideal membrane and the other three walls perfectly rigid.

We have used $\rho_{\mathrm{M}} = 5\,\mathrm{kg/m}^2$ and $T_0 = 2.5 \times 10^5\,\mathrm{N/m}$ as physical parameters for the membrane, whereas for the fluid we have used those of air: $\rho_{\mathrm{F}} = 1\,\mathrm{kg/m}^3$ and $c = 340\,\mathrm{m/s}$. Finally the cavity has been taken $1\,\mathrm{m}$ large and $0.5\,\mathrm{m}$ high.

Successive refinements of a basic uniform mesh have been used to determine orders of convergence. Table I shows the computed lowest vibration frequencies for both methods and each mesh; the parameter N denotes the number of elements in the fluid. We denote by ω_i^M the "membrane" eigenmodes and by ω_j^F the "fluid" ones. By "membrane" eigenmodes we mean those which are modes of the membrane perturbed by the interaction with the fluid, and by "fluid" eigenmodes those which are perturbations of the pure acoustic modes of the fluid in a rigid cavity. No exact solution of this problem is known in this case, but the orders of convergence, in powers of the meshsize h, have been estimated by extrapolation and are also included in the table.

An excelent performance can be observed for both methods. Moreover both give quite similar results. The computed order of convergence almost coincide with 2 as Theorem 1 predicts for Raviart-Thomas' elements.

Table I: Air-membrane coupled problem: computed eigenfrequencies

mode	Chen-Taylor				Raviart-Thomas			
	$N = 32$	$N = 128$	$N = 512$	order	$N = 32$	$N = 128$	$N = 512$	order
ω_1^M	720.42	717.13	716.31	2.00	720.52	717.16	716.32	2.00
ω_1^F	1056.20	1049.44	1047.74	1.99	1056.23	1049.45	1047.74	1.99
ω_2^M	1442.30	1417.13	1410.90	2.02	1443.23	1417.37	1410.96	2.01
ω_2^F	2108.42	2038.61	2020.48	1.94	2109.63	2039.01	2020.59	1.94
ω_3^F	2211.90	2170.43	2160.11	2.01	2211.94	2170.44	2160.11	2.01
ω_3^M	2289.50	2226.89	2212.09	2.08	2291.36	2227.24	2212.17	2.09
ω_4^F	2433.66	2409.94	2403.73	1.93	2457.99	2415.93	2405.22	1.97
ω_4^M	3059.36	2848.42	2795.64	1.99	3117.45	2849.75	2795.98	2.32

4.2 A pure acoustic problem on a curved domain

As a second test problem we have computed the natural vibration modes of a compressible fluid in a rigid 2D cavity with a curved boundary. The cavity is a circular sector of radius R and angle α. In this case, the analytical solutions of problem (1-5) are known; the exact vibration frequencies are $\omega_{nm} = c\,(x_{nm}/R)$, $n = 0, 1, 2, \ldots$, $m = 1, 2, \ldots$, where x_{nm} is the mth positive zero of the derivative of first kind Bessel function $J'_{n\alpha/\pi}(x)$.

We have used the physical parameters of air as in the previous test for the fluid and the following geometrical parameters: $R = 20\,\text{m}$, and $\alpha = \pi/2$.

Successive refinements of a basic mesh have been used again to determine orders of convergence and we also denote by N the number of elements of each mesh in the fluid. In Table II we show the exact lowest vibration frequencies and those computed with Raviart-Thomas' elements and Chen-Taylor's method with the same meshes. We also include, for each method, the computed order of convergence in powers of the meshsize.

Table II: Air in a rigid cavity with curved boundary: computed and exact eigenfrequencies.

mode	Chen-Taylor				Raviart-Thomas				exact
	$N{=}75$	$N{=}108$	$N{=}147$	order	$N{=}75$	$N{=}108$	$N{=}147$	order	
ω_{21}	51.09	51.18	51.25	0.48	52.39	52.25	52.16	2.02	51.92
ω_{02}	65.40	65.32	65.27	2.01	65.78	65.58	65.46	2.03	65.14
ω_{41}	87.86	88.13	88.36	0.60	92.24	91.68	91.34	2.00	90.40
ω_{22}	115.05	114.68	114.47	2.36	115.86	115.30	114.95	2.00	114.00
ω_{03}	121.15	120.57	120.22	1.91	121.69	120.95	120.50	1.94	119.26

We observe that the order of convergence is significatively lower for the eigenmodes ω_{21} and ω_{41} when they are computed with Chen-Taylor's method. This can be easily explained. In fact, for each mesh, the curved part of the boundary Γ is approximated by a polygonal Γ_h. Let K and K' be two quadrilaterals sharing a boundary node $P \in \Gamma$, with corresponding outward normal vectors n_h and n'_h, respectively, as shown in Figure 1.

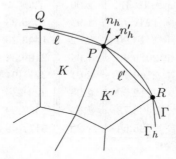

Figure 1: Mesh around a node on a curved boundary.

When the boundary condition (5) is imposed in each one of these quadrilaterals, we have $U_h(P) \cdot n_h = 0$ and $U_h(P) \cdot n'_h = 0$. Since, for a corner point P, n_h and n'_h are linearly independent, then we have $U_h(P) = 0$; that is, both components of U_h, the tangential and the normal one, vanish at each corner point of the boundary. Now, for a curved boundary Γ, all the nodes on Γ_h are generally corner points, then the tangential component of U_h should also vanish on the whole Γ_h. Because of this, when Chen-Taylor's method is used, the fluid displacement present a boundary layer near the curved boundary allowing this component to vanish. This can be observed in Figure 2 where the displacement fields corresponding to ω_{21} are shown as computed by both methods.

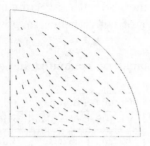
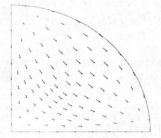

Chen-Taylor's method   Raviart-Thomas' elements

Figure 2: Displacement fields computed by each method for ω_{21}.

In our problem, ω_{21} and ω_{41} are the vibration frequencies in Table II, for which the corresponding displacement fields have non vanishing tangential components along the curved boundary. This is the reason why Chen-Taylor's method give poor approximations of these eigenfrequencies. Let us remark that a similar effect would be present in the case of a curved fluid-elastic solid interface.

To avoid this drawback, we may proceed as in the case of the interface condition for Raviart-Thomas' elements and impose (5) in the following weak form: if Γ_h is the discrete boundary associated with a curved part of Γ_R, for each edge $\ell \subset \Gamma_h$ with outward normal vector n_h, we impose $U_h \cdot n_h = 0$ only at the midpoint of ℓ. Since both components of $U_h|_\ell$ are linear this is equivalent to assume that $[U_h(P) + U_h(Q)] \cdot n_h = 0$ (using once more the notation of Figure 1).

In Table III we show the lowest vibration frequencies computed in this way and their orders of convergence which are now almost exactly 2 for all of them. Once more, we observe that the results almost coincide for both methods. The displacement fields computed in this way are practically identical to that of Figure 2 computed by using Raviart-Thomas' elements.

Table III: Chen-Taylor's method with weak kinematic conditions.

mode	$N = 75$	$N = 108$	$N = 147$	order	exact
ω_{21}	52.30	52.18	52.11	2.07	51.92
ω_{02}	65.40	65.32	65.27	2.02	65.14
ω_{41}	91.80	91.36	91.10	2.17	90.40
ω_{22}	115.40	114.97	114.71	2.04	114.00
ω_{03}	121.15	120.57	120.22	2.00	119.26

Finally, let us remark that such anomalous behavior in the vicinity of a curved edge is not exclusive of Chen-Taylor's method. In fact, it will be present for any discretization of a displacement formulation using nodal displacements as degrees of freedom. For instance, we have observed a similar effect when using the method in [6]. The technique proposed in this paper to correct such defect by imposing weakly the kinematic constraints is also valid in general.

5 Conclusions

Two methods to compute vibration modes of fluid-solid coupled problems have been considered. Both consist of discretizations of a pure displacement formulation of the problem. The first one is introduced in this paper for the case of meshes of quadrilaterals. It is based on Raviart-Thomas' elements for the fluid coupled in a nonconforming way with the elements used for the solid. The second one is Chen-Taylor's method which is based on standard bilinear elements for the fluid coupled in a direct conforming way with the solid.

For the first one, rigorous error estimates and a mathematical proof that spurious modes cannot arise are valid. The second one is based only on heuristic considerations. However, in the case of a polygonal fluid domain, the latter is much simpler to be dealt with. The results obtained in all the considered problems with both methods are essentially the same.

For fluid domains with curved boundaries, direct application of Chen-Taylor's method still converges but the order of convergence deteriorates. A methodology is proposed to recover optimal orders of convergence. The quality of the approximation in this case is shown to be the same as for polygonal domains.

References

[1] BERMÚDEZ, A., DURÁN, R., MUSCHIETTI, M.A., RODRÍGUEZ, R. AND SOLOMIN, J., *Finite element vibration analysis of fluid-solid systems without spurious modes*, SIAM J. Numer. Anal., 32 (1995) 1280-1295.

[2] BERMÚDEZ, A., DURÁN, R. AND RODRÍGUEZ, R., *Finite element analysis of compressible and incompressible fluid-solid systems*, Math. Comp. (to appear).

[3] BERMÚDEZ, A., DURÁN, R. AND RODRÍGUEZ, R., *Finite element solution of incompressible fluid-structure vibration problems*, Int. J. Numer. Methods Eng., 40 (1997) 1435-1448.

[4] BERMÚDEZ, A. AND RODRÍGUEZ, R., *Finite element computation of the vibration modes of a fluid-solid system*, Comp. Methods in Appl. Mech. and Eng., 119 (1994) 355-370.

[5] CHEN, H.C. AND TAYLOR, R.L., *Vibration analysis of fluid-solid systems using a finite element displacement formulation*, Int. J. Numer. Methods Eng., 29 (1990) 683-698.

[6] HAMDI, M., OUSET, Y. AND VERCHERY, A., *A displacement method for the analysis of vibrations of coupled fluid-structure systems*, Int. J. Numer. Methods Eng., 13 (1978) 139-150.

[7] MORAND, H.J-P. AND OHAYON, R., *Fluid Structure Interaction*, J. Wiley & Sons, 1992.

[8] RAVIART, P.A. AND THOMAS, J.M., *A mixed finite element method for second order elliptic problems*, in *Mathematical Aspects of Finite Element Methods*, Lecture Notes in Mathematics 606, Springer Verlag, 1972, pp. 292-315.

[9] RODRÍGUEZ, R. AND SOLOMIN, J., *The order of convergence of eigenfrequencies in finite element approximations of fluid-structure interaction problems*, Math. Comp., 65 (1996) 1463-1475.

LEOPOLDO P. FRANCA

An Overview of the Residual-Free-Bubbles Method

1 Introduction

Stabilized methods and other nonstandard techniques of discretization were developed to deal with intricate physical problems governed by singularly perturbed equations, and/or difficulties in approximating system of equations, etc. The idea then was to develop methods that would not completely change the structure of a simple finite element code based on piecewise polynomials of equal order approximation for all variables. Surprisingly the idea is effective and can be applied to a variety of applications (see [5-6, 8-11, 16-19] and references therein).

Wishing to further understand stabilized finite element methods and other nonstandard Galerkin finite element methods, we have revisited the Galerkin method using richer subspaces other than piecewise polynomials. The idea is to enlarge the space of piecewise polynomials with functions defined elementwise, such that improved accuracy and stability are achieved, which are also goals shared by stabilized methods. Noting first that streamline diffusion can be obtained by this process [2], a theory was developed to show that *virtual* bubbles can be constructed to reproduce stabilized methods in a variety of applications and based on piecewise polynomials of all orders [1]. The open question then was: Is there a systematic procedure to construct these virtual bubbles so that improved discretizations are developed regardless of how those may look at the end? In other words, dropping the requirement to obtain a stabilized method, let us pursue ultimately accurate and stable methods.

This paper deals with this question: we select a space of bubble functions that are spanned by the exact solution minus its piecewise linear contribution on an element. This is the space of residual-free-bubbles that were suggested in [4] and crystallized in [12-14]. An approach similar to the latter one was developed independently in [15] motivated by physical arguments and the equivalence to the residual-free-bubbles idea is presented in [3].

We present the residual-free-bubbles idea and some of its applications in the next Section. In particular we examine the consequences of this choice of discretization for the Timoshenko beam problem and for a heat conduction application. It is interesting to observe that old numerical tricks, such as selective reduced integration and mass lumping, naturally arises from the choice of this bubble space of functions.

Another family of methods suggested by static condensation, closely related to stabilized methods, is presented in [7] and [20]. Therein we *fix* the space of bubbles to be spanned by residual-free bubbles and explore the methods unveiled by static condensation of these residual-free bubbles.

2 Residual Free Bubbles

This seems to be a very promising approach in that a systematic derivation of methods is now possible. To define residual free bubbles let us consider the standard Galerkin method for

$$Lu = f \qquad \text{in } \Omega$$

$$u = 0 \qquad \text{on } \Gamma = \partial\Omega$$

(1)

where L is a linear differential operator and f is a given function. Then, we wish to find u_h such that

$$a(u_h, v_h) = (Lu_h, v_h) = (f, v_h) \qquad \forall v_h$$

(2)

Here u_h and v_h are piecewise polynomials plus bubble functions, i.e.,

$$u_h = u_1 + u_b$$

(3)

where the bubble functions satisfy the differential equations strongly, i.e.,

$$Lu_b = -(Lu_1 - f) \qquad \text{in } K$$

(4)

subject to zero Dirichlet boundary condition on the element boundary, i.e.,

$$u_b = 0 \qquad \text{on } \partial K .$$

(5)

Problem given by equations (4)-(5) are addressed by solving instead:

$$L\varphi_{i,K} = -L\psi_{i,K} \qquad \text{in } K$$

(6)

$$\varphi_{i,K} = 0 \qquad \text{on } \partial K$$

(7)

where the $\psi_{i,K}$'s are the local basis functions for u_1 and

$$L\varphi_{f,K} = f \qquad \text{in } K$$

(8)

$$\varphi_{f,K} = 0 \qquad \text{on } \partial K .$$

(9)

Thus, if $u_{1|K} = \sum_{i=1}^{n_{en}} c_{i,K}\psi_{i,K}$ then

$$u_{b|K} = \sum_{i=1}^{n_{en}} c_{i,K}\varphi_{i,K} + \varphi_{f,K}$$

(10)

with the same coefficients $c_{i,K}$'s.

84

We now wish to address the question: what does the method (2) imply for the reduced space of polynomials? (or, what is the effect of u_b on the u_1 part of the solution?)

The answer is to use *static condensation* (as before): first $v = v_{b,K}$ on K (zero elsewhere):

$$a(u_1 + u_b, v_{b,K})_K = (f, v_{b,K})_K . \tag{11}$$

But this equation is satisfied automatically due to our choice of bubbles. Indeed this equation is the variational equation for

$$Lu_b = -(Lu_1 - f) \qquad \text{in } K \tag{12}$$

using $v = v_{b,K}$ on K (zero elsewhere) as test functions.

The second part of static condensation is: use $v = v_1$ in (2):

$$a(u_1 + u_b, v_1) = (f, v_1)$$

$$a(u_1, v_1) + a(u_b, v_1) = (f, v_1) \tag{13}$$

This is the method suggested by static condensation. Using residual free bubbles the second term modification due to the bubbles is computed after solving equations (6)-(9).

Let us consider two examples of this approach to show that mass lumping and selective reduced integration are tricks that can be 'explained' by residual-free-bubbles.

The presentation now follows [13]. The first example is: Find a scalar valued function $u(x)$ defined in $\Omega \subset \mathbb{R}$ such that

$$\sigma u - \kappa u'' = f \qquad \text{in } \Omega \tag{14}$$

$$u = 0 \qquad \text{on } \Gamma = \partial\Omega \tag{15}$$

where σ and κ are given positive constants and $f(x)$ is a given source function.

Here:

$$Lu = \sigma u - \kappa u''$$

and the residual free bubble problems given by eqs. (6)-(9) are:

$$\sigma \varphi_{i,K} - \kappa \varphi_{i,K}'' = -\sigma \psi_{i,K} \qquad \text{in } K \tag{16}$$

$$\varphi_{i,K} = 0 \qquad \text{on } \partial K \tag{17}$$

where the $\psi_{i,K}$'s are the basis functions for u_1 and have second derivative zero inside each element and

$$\sigma\varphi_{f,K} - \kappa\varphi''_{f,K} = f \qquad \text{in } K \tag{18}$$

$$\varphi_{f,K} = 0 \qquad \text{on } \partial K. \tag{19}$$

The solutions of (16)-(19) with respect to the local coordinate $\xi \in [0, h_K]$ are:

$$\varphi_{1,K}(\xi) = \frac{\sinh\left(\sqrt{\frac{\sigma}{\kappa}}(h_K - \xi)\right)}{\sinh\left(\sqrt{\frac{\sigma}{\kappa}}h_K\right)} - \left(1 - \frac{\xi}{h_K}\right)$$

$$\varphi_{2,K}(\xi) = \frac{\sinh\left(\sqrt{\frac{\sigma}{\kappa}}\xi\right)}{\sinh\left(\sqrt{\frac{\sigma}{\kappa}}h_K\right)} - \frac{\xi}{h_K} \tag{20}$$

$$\varphi_{f,K}(\xi) = -\frac{f}{\sigma}[\varphi_{1,K}(\xi) + \varphi_{2,K}(\xi)]$$

for piecewise-constant loads.

The Galerkin method given by the second step of static condensation is

$$a(u_1, v_1) + a(u_b, v_1) = (f, v_1) \tag{21}$$

and in this case

$$(\sigma u_1, v_1) + (\kappa u'_1, v'_1) + \sum_K (\sigma u_{b,K}, v_1)_K = (f, v_1) \tag{22}$$

If we substitute the expression for $u_{b|K} = \sum_{i=1}^{n_{en}} c_{i,K}\varphi_{i,K} + \varphi_{f,K}$ with the exact results above then we are led to a system of equations for the unknowns constants $c_{i,K}$'s.

If we write the system of equations for a uniform mesh, then a typical interior node satisfies the following: (after some algebra)

$$A_h \left[\frac{-c_{I-1} + 2c_I - c_{I+1}}{h} \right] + B_h c_I = C_h f \tag{23}$$

where

$$A_h = \kappa \frac{\left(\sqrt{\frac{\sigma}{\kappa}}h_K\right)}{\sinh\left(\sqrt{\frac{\sigma}{\kappa}}h_K\right)} \tag{24}$$

$$B_h = 2\sqrt{\sigma\kappa} \tanh\left(\frac{1}{2}\sqrt{\frac{\sigma}{\kappa}}h_K\right) \tag{25}$$

$$C_h = \frac{1}{\sigma}B_h. \tag{26}$$

86

This is the method implied by the residual-free bubbles approach. The method will give *nodal exact values* for all κ, σ, f and h. For small $\sqrt{\frac{\sigma}{\kappa}} h_K$ we have

$$A_h \approx \kappa, \quad B_h \approx \sigma h, \quad C_h \approx h \tag{27}$$

simplifying the method to:

$$\kappa \left[\frac{-c_{I-1} + 2c_I - c_{I+1}}{h} \right] + \sigma h c_I = hf \tag{28}$$

which is form-identical to the equations produced by the standard Galerkin method using piecewise linears with full integration on the second derivative term and 'mass lumping' in the zero order term.

The presentation now follows [14]. The second example is developed to show the appearance of selective reduced integration from residual-free-bubbles. The Timoshenko beam model is governed by the following differential equations (after non-dimensionalization):

$$-\theta'' - \frac{1}{\epsilon^2}(w' - \theta) = 0 \quad \text{in } \Omega$$
$$-\frac{1}{\epsilon^2}(w'' - \theta') = f \quad \text{in } \Omega \tag{29}$$

where prime denotes differentiation with respect to $x \in \Omega = (0, 1)$, θ and w are the rotation and displacement variables, f is the load and ϵ is a non-dimensional parameter proportional to the beam thickness.

To (29) we append the following clamped boundary conditions (other boundary conditions may be used without major changes in what follows):

$$w(0) = w(1) = 0$$
$$\theta(0) = \theta(1) = 0. \tag{30}$$

The variational formulation corresponding to (29)-(30) is given by: Find $\{\theta, w\} \in H_0^1(\Omega)^2$ such that

$$(\theta', \psi') + \frac{1}{\epsilon^2}(w' - \theta, v' - \psi) = (f, v) \qquad \forall \{\psi, v\} \in H_0^1(\Omega)^2 \tag{31}$$

where we use the notation $(f, g) = \int_\Omega fg \, d\Omega$.

Consider a partition of Ω into non-overlapping elements in the usual way. Then the exact solution of our problem can be decomposed into:

$$\theta = \theta_1 + \theta_b$$
$$w = w_1 + w_b \tag{32}$$

87

where θ_1 and w_1 are spanned by the standard continuous piecewise linears of finite element methods, and θ_b and w_b are assumed to satisfy the following differential equations in each element K:

$$-\theta_b'' - \frac{1}{\epsilon^2}(w_b' - \theta_b) = -\left(-\theta_1'' - \frac{1}{\epsilon^2}(w_1' - \theta_1)\right)$$

$$-\frac{1}{\epsilon^2}(w_b'' - \theta_b') = -\left(-\frac{1}{\epsilon^2}(w_1'' - \theta_1') - f\right)$$

(33)

and subjected to the boundary conditions:

$$\theta_b = w_b = 0 \qquad \text{on } \partial K .$$

(34)

Equations (33) can be rewritten as (note that $\theta_1'' = w_1'' = 0$ in K):

$$-\epsilon^2\theta_b'' + \theta_b - w_b' = w_1' - \theta_1$$

$$\theta_b' - w_b'' = -\theta_1' + \epsilon^2 f .$$

(35)

¿From $(35)_1$

$$\theta_b - w_b' = w_1' - \theta_1 + \epsilon^2\theta_b''$$

and combining with $(35)_2$ we get

$$\theta_b''' = f \qquad \text{in } K .$$

(36)

Integrating three times (with respect to the local variable in the element, $\xi \in [0, h_K], h_K = x_{i+1} - x_i, \xi = x - x_i$) and assuming piecewise constant load f, and for notation's sake dropping the subscripts for h and f (nowhere we need to assume that h_K is constant in what follows) we get:

$$\theta_b(\xi) = \frac{\xi^3}{6}f + c_1\frac{\xi^2}{2} + c_2\xi + c_3 .$$

(37)

Applying the boundary conditions $\theta_b(0) = \theta_b(h) = 0$ above gives:

$$\theta_b(\xi) = \frac{\xi}{6}f(\xi^2 - h^2) + c_1\frac{\xi}{2}(\xi - h) .$$

(38)

Using this expression into the first equation of (35) after one integration we get:

$$w_b(\xi) = \int_0^\xi \theta_1(t)\, dt - w_1(\xi) - \epsilon^2\left[\frac{f}{6}(3\xi^2 - h^2) + \frac{c_1}{2}(2\xi - h)\right]$$

$$+ \frac{f}{6}\left[\frac{\xi^4}{4} - \frac{\xi^2}{2}h^2\right] - \frac{c_1}{12}\xi^2(3h - 2\xi) + c_4 .$$

(39)

Applying the boundary conditions $w_b(0) = w_b(h) = 0$ in (39) we get expressions for the remaining constants c_1 and c_4 and the expressions for the residual-free bubble functions are then given by:

$$\theta_b(\xi) = f\left\{\frac{\xi}{6}(\xi^2 - h^2) + \frac{h\xi}{4}(h - \xi)\right\}$$

$$+ \frac{1}{\epsilon^2 + \frac{h^2}{12}}\frac{\xi(\xi - h)}{2}\left\{\theta_1(\frac{h}{2}) - \frac{w_1(h) - w_1(0)}{h}\right\}, \tag{40}$$

$$w_b(\xi) = \xi\left(1 - \frac{\xi}{2h}\right)\theta_1(0) + \frac{\xi^2}{2h}\theta_1(h) + \frac{\xi}{h}[w_1(0) - w_1(h)]$$

$$- \xi\left[\epsilon^2 - \frac{\xi^2}{6} + \frac{h\xi}{4}\right]\left\{\frac{1}{\epsilon^2 + \frac{h^2}{12}}\left[\theta_1(\frac{h}{2}) - \frac{w_1(h) - w_1(0)}{h}\right] - \frac{hf}{2}\right\} \tag{41}$$

$$+ \frac{f\xi^2}{2}\left[-\epsilon^2 + \frac{\xi^2}{12} - \frac{h^2}{6}\right].$$

If we take the test functions $\psi = \psi_1$ and $v = v_1$, where ψ_1 and v_1 are spanned by continuous piecewise linears, then using decomposition (32) the variational formulation (31) can be rewritten as

$$(\theta_1', \psi_1') + \frac{1}{\epsilon^2}(w_1' - \theta_1, v_1' - \psi_1) - (f, v_1) + \frac{1}{\epsilon^2}(w_b' - \theta_b, v_1' - \psi_1) = 0 \tag{42}$$

where, by integration-by-parts, we used that:

$$(\theta_b', \psi_1') = \sum_K(\theta_b', \psi_1')_K = \sum_K[(\theta_b, \psi_1')_{\partial K} - (\theta_b, \psi_1'')_K] = 0. \tag{43}$$

Note that (42) consists of the Galerkin method for equal-order piecewise linear approximations for θ and w (without tricks, using full integration) plus a "perturbation term" that we need to compute based on the bubble functions given by (40) and (41). First by (40) and (41) we compute:

$$w_b' - \theta_b = \theta_1(0) + \frac{\xi}{h}[\theta_1(h) - \theta_1(0)] - \frac{w_1(h) - w_1(0)}{h}$$

$$- \frac{\epsilon^2}{\epsilon^2 + \frac{h^2}{12}}\left[\theta_1(\frac{h}{2}) - \frac{w_1(h) - w_1(0)}{h}\right] + \epsilon^2 f\left(\frac{h}{2} - \xi\right). \tag{44}$$

Note also that

$$w_1' - \theta_1 = \frac{w_1(h) - w_1(0)}{h} - \left(1 - \frac{\xi}{h}\right)\theta_1(0) - \frac{\xi}{h}\theta_1(h). \tag{45}$$

Thus summing (44) to (45)

$$w_1' - \theta_1 + w_b' - \theta_b = \epsilon^2 f\left(\frac{h}{2} - \xi\right) - \frac{\epsilon^2}{\epsilon^2 + \frac{h^2}{12}}\left[\theta_1(\frac{h}{2}) - \frac{w_1(h) - w_1(0)}{h}\right]. \qquad (46)$$

Therefore, using (46), the variational formulation given by (42) reduces to

$$(\theta_1', \psi_1') + \sum_K \frac{1}{\epsilon^2 + \frac{h_K^2}{12}}\left(\frac{w_1(h_K) - w_1(0)}{h_K} - \theta_1(\frac{h_K}{2}), v_1' - \psi_1\right)_K$$

$$= (f, v_1) + \sum_K f_K(\xi - \frac{h_K}{2}, v_1' - \psi_1)_K \qquad (47)$$

where we reintroduced the subscripts for h and the piecewise constant load f. This can also be rewritten as

$$(\theta_1', \psi_1') + \sum_K \frac{1}{\epsilon^2 + \frac{h_K^2}{12}}(w_1' - R\theta_1, v_1' - \psi_1)_K = (f, v_1)$$

$$+ \sum_K f_K(\xi - \frac{h_K}{2}, v_1' - \psi_1)_K \qquad (48)$$

where R stands for a reduced integration operator.

Formulation (48) was *derived* using full integration throughout and by construction its solution is nodally exact. The final form is identical to applying the following tricks to the standard variational formulation:

  i) Use one-point reduced integration on the shear energy term;
 ii) Replace its coefficient $1/\epsilon^2$ by $1/(\epsilon^2 + (h_K^2/12))$ in each element;
iii) Correct the right-hand-side as in equation (48) for piecewise-constant loads.

To emerge with these collection of "tricks" requires ingenuity and for the first two tricks different arguments have been given before by several authors.

Acknowledgment. The author acknowledges the support by the Chilean CONI-CYT to present this paper in Concepción, Chile, in December of 1995.

References

[1] C. BAIOCCHI, F. BREZZI AND L.P. FRANCA, "Virtual bubbles and the Galerkin-least-squares method", *Computer Methods in Applied Mechanics and Engineering*, Vol.**105** (1993) 125-141.

[2] F. BREZZI, M.O. BRISTEAU, L.P. FRANCA, M. MALLET AND G. ROGE, "A relationship between stabilized finite element methods and the Galerkin method with bubble functions," *Computer Methods in Applied Mechanics and Engineering*, Vol.**96** (1992) 117-129.

[3] F. BREZZI, L.P. FRANCA, T.J.R. HUGHES AND A. RUSSO, "$b = \int g$," *Computer Methods in Applied Mechanics and Engineering*, to appear.

[4] F. BREZZI AND A. RUSSO, "Choosing bubbles for advection-diffusion problems," *Math. Models Meths. Appl. Sci.* Vol.**4** (1994) 571-587.

[5] A.N. BROOKS AND T.J.R. HUGHES, "Streamline upwind/Petrov-Galerkin formulations for convective dominated flows with particular emphasis on the incompressible Navier-Stokes equations," *Computer Methods in Applied Mechanics and Engineering*, Vol.**32** (1982) 199-259.

[6] L.P. FRANCA AND E.G.DUTRA DO CARMO, "The Galerkin gradient least-squares method," *Computer Methods in Applied Mechanics and Engineering*, Vol.**74** (1989) 41-54.

[7] L.P. FRANCA AND C. FARHAT, "Bubble functions prompt unusual stabilized finite element methods," *Computer Methods in Applied Mechanics and Engineering*, Vol.**123**, pp.299-308 (1995).

[8] L.P. FRANCA AND S.L. FREY, "Stabilized finite element methods: II. The incompressible Navier-Stokes Equations", *Comput. Methods Appl. Mech. Engrg.*, Vol.**99** (1992) 209-233.

[9] L.P. FRANCA, S.L. FREY AND T.J.R. HUGHES, "Stabilized finite element methods: I. Application to the advective-diffusive model", *Comput. Methods Appl. Mech. Engrg.* Vol.**95** (1992) 253-276.

[10] L.P. FRANCA AND T.J.R. HUGHES, "Two classes of mixed finite element methods", *Comput. Methods Appl. Mech. Engrg.* Vol.**69** (1988) 89-129.

[11] L.P. FRANCA, T.J.R. HUGHES AND R. STENBERG, "Stabilized finite element methods for the Stokes problem", pp. 87-107 in *Incompressible Computational Fluid Dynamics-Trends and Advances*, M.D. Gunzburger and R.A. Nicolaides eds., Cambridge University Press, 1993.

[12] L.P. FRANCA AND A. RUSSO, "Deriving upwinding, mass lumping and selective reduced integration by residual-free bubbles," *Applied Mathematics Letters*, Vol.**9** (1996) 83-88.

[13] L.P. FRANCA AND A. RUSSO, "Mass lumping emanating from residual-free bubbles," *Computer Methods in Applied Mechanics and Engineering*, Vol.**142** (1997) 353-360.

[14] L.P. FRANCA AND A. RUSSO, "Unlocking with residual-free bubbles," *Computer Methods in Applied Mechanics and Engineering*, Vol.**142** (1997) 361-364.

[15] T.J.R. HUGHES, "Multiscale phenomena: Green's functions, the Dirichlet-to-Neumann formulation, subgrid scale models, bubbles and the origin of stabilized methods," *Computer Methods in Applied Mechanics and Engineering*, Vol.**127** (1995) 387-401.

[16] T.J.R. HUGHES AND L.P. FRANCA, "A new finite element formulation for computational fluid dynamics: VII. The Stokes problem with various well-posed boundary conditions: symmetric formulations that converge for all velocity/pressure spaces", *Comput. Methods Appl. Mech. Engrg.* Vol.**65** (1987) 85-96.

[17] T.J.R. HUGHES, L.P. FRANCA AND M. BALESTRA, "A new finite element formulation for computational fluid dynamics: V. Circumventing the Babuška-Brezzi condition: A stable Petrov-Galerkin formulation of the Stokes problem accommodating equal-order interpolations", *Comput. Methods Appl. Mech. Engrg.* Vol.**59** (1986) 85-99.

[18] T.J.R. HUGHES, L.P. FRANCA AND G.M. HULBERT, "A new finite element formulation for computational fluid dynamics: VIII. The Galerkin-least-squares method for advective-diffusive equations", *Comput. Methods Appl. Mech. Engrg.*, Vol.**73** (1989) 173-189.

[19] C. JOHNSON, U. NÄVERT AND J. PITKÄRANTA, "Finite element methods for linear hyperbolic problem," *Comput. Methods Appl. Mech. Engrg.*, Vol.**45** (1984) 285-312.

[20] M. LESOINNE, C.FARHAT AND L.P. FRANCA, "Unusual stabilized finite element methods for second order linear differential equations," In Part I, pp. 377-386 of the *Proceedings of the Ninth International Conference on Finite Elements in Fluids - New Trends and Applications*, (M. Morandi Cecchi, K. Morgan, J. Periaux, B.A. Schrefler, O.C. Zienkiewicz, eds.), Venice, Italy, October 1995.

SERGIO IDELSOHN and EUGENIO OÑATE

The finite point method (FPM) in computational mechanics

1 Introduction

It is widely acknowledged that 3D mesh generation remains one of the big challenges in both Finite Element (FE) [1] and Finite Volume (FV) [2-4] computations. Thus, given enough computer power even the most complex problems in computational mechanics, such as the 3D solution of Navier-Stokes equations in fluid flow can be tackled accurately providing an acceptable mesh is available. The generation of 3D meshes, however, is despite major recent advances in this field, certainly the bottle neck in most industrial FE and FV computations and, in many cases, it can absorb far more time and cost than the numerical solution itself.

Different authors have recently investigated the possibility of deriving numerical methods where meshes are unnecessary. The first attempts were reported by some finite difference (FD) practitioners deriving FD schemes in arbitrary irregular grids [5-8]. Here typically the concept of "star" of nodes was introduced to derive FD approximations for each central node by means of local Taylor series expansions using the information provided by the number and position of nodes contained in each star [9].

An alternative class of methods named Smooth Particle Hydrodynamics (SPH), sometimes called the Free Lagrange methods, depend only on a set of disordered point or particles and has enjoyed considerable popularity in computational physics and astrophysics to model the motion and collision of stars [9].

Nayroles et al. [10] proposed a technique which they call the Diffuse Element (DE) method, where only a collection of nodes and a boundary description is needed to formulate the Galerkin equations. The interpolating fuctions are polynomials fitted to the nodal values by a weighted least squares (WLS) approximation. Although no finite element mesh is explicitly required in this method, still some kind of "auxiliary grid" was used in [10] in order to compute numericaly the integral expressions derived from the Galerkin approach. Belytschko et al. [11,12] have proposed an extension of the DE approach which they call the element-free Galerkin (EFG) method. This provides additional terms in the derivatives of the interpolant considered unnecessary by Nayroles et al. [10]. In addition, a regular cell structure is chosen as the "auxiliary grid" to compute the integrals by means of high order quadratures. Duarte and Oden [13], Babuška and Melenk [14] and Taylor et al. [15] have recently formalized this type of approximation as a subclass of the so called "partition of unity" (PU) methods and they propose meshless and enhanced FE procedures using hierarchical PU interpolations.

Liu et al. [16-19] have developed a different class of "gridless" multiple scale methods based on reproducing kernel and wavelet analysis. This technique termed

Reproducing Kernel Particle (RKP) method introduces a new type of shape functions using an integral window transform. The window function can be *translated* and *dilated* around the domain thus replacing the need to define elements and providing refinement.

In a recent work Oñate *et al.* [20–23] have found that the weighted least square interpolation with a simple point collocation technique for evaluating the approximation integrals is a promising Finite Point Method (FPM) for the numerical solution of a wide range of problems in computational mechanics. The advantages of this FPM compared with standard FEM is to avoid the necessity of mesh generation and compared with classical FDM is the facility to handle the boundary conditions and the non-structured distribution of points. Moreover the FPM proposed seems to be as accurate as other numerical methods and the computing time to solve the differential equation is of the same order as for FE and FV methods using non-structured grids. Examples of application of the FPM to the solution of elliptic equations as well as to those governing convective transport and compressible fluid flow using linear WLS interpolations were reported in [20–23].

Another interesting conclusion reported in [22] is the comparison with other meshless techniques presented in the literature. Firstly, the use of a Gaussian weighting function improves considerably the results with respect to the standard least square (LSQ) approach. Secondly, the sensitivity of any point data interpolation based procedure to a variable number of points in each interpolation domain (cloud) must be low enough to preserve the freedom of adding, moving or removing points. This sensitivity is very high in meshless techniques using the LSQ approximation, it is large in WLS methods with linear interpolations and it quite low in WLS methods using quadratic interpolations which indicates some advantage of the later for practical applications [22].

In this paper the FPM proposed in [20–23] is further extended to the solution of the advective-convective transport equations as well as those governing the flow of compressible fluids using a *quadratic WLS interpolation*. Here the stabilization of the numerical algorithm is crucial to guarantee acceptable results similarly as it occurs in FD, FV and FE methods for fluid flow problems. A residual stabilization procedure, adequate for the FPM, is proposed in the paper. It is shown that the stabilization of both the convective terms and the Neumann boundary condition is necessary to ensure a correct solution in all cases.

In next section some basic concepts of mesh free techniques including some details of the different weighted least square interpolations typically used are briefly described.

2 Basic concepts of mesh free techniques

Let us assume a scalar problem governed by a differential equation

$$A(u) = b \quad \text{in} \quad \Omega \tag{1}$$

with Neumann and Dirichlet boundary conditions:

$$B(u) = t \quad \text{in} \quad \Gamma_t \quad ; \quad u - u_p = 0 \quad \text{in} \quad \Gamma_u \tag{2}$$

to be satisfied in a domain Ω with boundary $\Gamma = \Gamma_t \cup \Gamma_u$. In the above A and B are appropriate differential operators, u is the problem unknown and b and t represent external forces or sources acting over the domain Ω and along the boundary Γ_t, respectively. Finally u_p is the prescribed value of u over the boundary Γ_u.

The most general procedure of solving numerically the above system of differential equations is the weighted residual method in which the unknown function u is approximated by some trial approximation $\hat{u}$ and eqs.(1) and (2) are replaced by [1]

$$\int_{\Omega} W_i[A(\hat{u}) - b]\, d\Omega + \int_{\Gamma_t} \bar{W}_i[B(\hat{u}) - t]\, d\Gamma + \int_{\Gamma_u} \bar{\bar{W}}_i[\hat{u} - u_p]d\Gamma = 0 \tag{3}$$

with the weighting functions W_i, $\bar{W}_i$ and $\bar{\bar{W}}_i$ defined in different ways FE, FV and FD methods can be considered as particular cases of (3) and indeed so can all the meshless approximation procedures.

In order to keep a local character of the problem (leading to a banded matrix), function u must be approximated by a combination of locally defined functions as

$$u(x) \cong \hat{u}(x) = \sum_{i=1}^{n_p} N_i(x)u_i^h = \mathbf{N}^T(x)\mathbf{u}^h \tag{4}$$

with n_p being the total number of points in the domain and the interpolation functions $N_i(x)$ satisfy

$$N_i(x) \neq 0 \quad \text{if} \quad x \in \Omega_i \quad ; \quad N_i(x) = 0 \quad \text{if} \quad x \notin \Omega_i \tag{5}$$

Here Ω_i is a subdomain of Ω containing n points, $n \ll n_p$. In (4) u_i^h is the *approximate value* of u at point i such that $u(x_i) \simeq u_i^h$.

In FE and FV methods the Ω_i subdomains are divided into elements and the N_i function may have some discontinuities (in the function itself or in its derivatives) on the element interfaces. In the FE method the weighting functions W_i are defined in "weighting domains" which usually coincide precisely with the interpolating domains Ω_i. In cell vertex FV the interpolation and integration domains also coincide, however in the cell centered case they are different [2,3,4,22].

A common feature of FE and FV methods is that they both require a mesh for interpolation purposes and also to compute the integrals in eq. (3). FE methods define the shape functions N_i over non overlapping regions (elements) the assembly of which constitues the domain Ω_i [1]. Different interpolations are therefore possible for a given number of points simply by changing the orientation or the form of these regions. Although FV techniques do not explicitly define an interpolation of the form (4), it is well known that they are equivalent to using linear shape functions over domains Ω_i defined in the same manner as in the FE method [2-4].

The way to define the shape functions N_i is classical in the FEM [1]. The process will be repeated here in order to compare and better undertand the differences with the FPM.

Let Ω_i be the interpolation domain (cloud) of a function $u(x)$ and let s_j with $j = 1, 2, \cdots, n$ be a collection of n points with coordinates $x_j \in \Omega_i$. The unknown function u may be approximated within Ω_i by

$$u(x) \cong \hat{u}(x) = \sum_{l=1}^{m} p_l(x)\alpha_l = \mathbf{p}(x)^T \boldsymbol{\alpha} \qquad (6)$$

where $\boldsymbol{\alpha} = [\alpha_1, \alpha_2, \cdots \alpha_m]^T$ and vector $\mathbf{p}(x)$ contains typically monomials, hereafter termed "base interpolating functions", in the space coordinates ensuring that the basis is complete. For a 2D problem we can specify

$$\mathbf{p} = [1, x, y]^T \quad \text{for} \ m = 3 \quad ; \quad \mathbf{p} = [1, x, y, x^2, xy, y^2]^T \quad \text{for} \ m = 6 \quad \text{etc. (7)}$$

Function $u(x)$ can now be sampled at the n points belonging to Ω_i giving

$$\mathbf{u}^h = \left\{ \begin{array}{c} u_1^h \\ u_2^h \\ \vdots \\ u_n^h \end{array} \right\} \cong \left\{ \begin{array}{c} \hat{u}_1 \\ \hat{u}_2 \\ \vdots \\ \hat{u}_n \end{array} \right\} = \left\{ \begin{array}{c} \mathbf{p}_1^T \\ \mathbf{p}_2^T \\ \vdots \\ \mathbf{p}_n^T \end{array} \right\} \boldsymbol{\alpha} = \mathbf{C}\boldsymbol{\alpha} \qquad (8)$$

where $u_j^h = u(x_j)$ are the unknown but sought for values of function u at point j, $\hat{u}_j = \hat{u}(x_j)$ are the approximate values, and $\mathbf{p}_j = \mathbf{p}(x_j)$.

In the FE approximation the number of points is chosen so that $m = n$. In this case $\mathbf{C}$ is a square matrix and we can obtain *after equaling u^h with $\mathbf{C}\boldsymbol{\alpha}$* in (8)

$$\boldsymbol{\alpha} = \mathbf{C}^{-1}\mathbf{u}^h \qquad (9)$$

and

$$u \cong \hat{u} = \mathbf{p}^T \mathbf{C}^{-1}\mathbf{u}^h = \mathbf{N}^T \mathbf{u}^h = \sum_{j=1}^{n} N_j^i u_j^h \qquad (10)$$

with $\mathbf{N}^T = [N_1^i, \cdots, N_n^i] = \mathbf{p}^T \mathbf{C}^{-1}$ and $N_j^i = \sum_{l=1}^{m} p_l(x)C_{lj}^{-1} \qquad (11)$

The shape functions $N_j^i(x)$ satisfy the standard condition [1]

$$\begin{aligned} N_j^i(x_i) &= 1 \quad j = i \\ &= 0 \quad j \neq i \qquad i, j = 1, \cdots, n \end{aligned} \qquad (12)$$

Furthermore, for two differents interpolating domains Ω_i and Ω_k the corresponding shape functions are the same, i.e.: $N_j^i = N_j^k$

The development of the N_j^i can often be performed directly using interpolation methods and/or isoparametric concepts [1].

If $n > m$, $\mathbf{C}$ is no longer a square matrix and the approximation can not fit all the u_j^h values. This problem can be simply overcome by determining the $\hat{u}$ values by minimizing the sum of the square distances of the error at each point weighted with a function $\varphi(x)$ as

$$J = \sum_{j=1}^{n} \varphi(x_j) \left(u_j^h - \hat{u}(x_j) \right)^2 = \sum_{j=1}^{n} \varphi(x_j) \left(u_j^h - \mathbf{p}_j^T \boldsymbol{\alpha} \right)^2 \qquad (13)$$

with respect to the $\boldsymbol{\alpha}$ parameters. Note that for $\varphi(x) = 1$ the standard least square (LSQ) method is reproduced.

Function $\varphi(x)$ is usually built in such a way that it takes a unit value in the vicinity of the point i (typically called "star node" [8]) where the function (or its derivatives) are to be computed and vanishes outside a region Ω_i surrounding the point. The region Ω_i can be used to define the number of sampling points n in the interpolation region. A typical choice for $\varphi(x)$ is the normalized Gaussian function. Of course $n \geq m$ is always required in the sampling region and if equality occurs no effect of weighting is present and the interpolation is the same as in the LSQ scheme.

Standard minimization of eq.(13) with respect to $\boldsymbol{\alpha}$ gives

$$\boldsymbol{\alpha} = \bar{\mathbf{C}}^{-1} \mathbf{u}^h \quad , \quad \bar{\mathbf{C}}^{-1} = \mathbf{A}^{-1} \mathbf{B} \qquad (14)$$

with matrices $\mathbf{A}$ and $\mathbf{B}$ given by

$$\mathbf{A} = \sum_{j=1}^{n} \varphi(x_j) \mathbf{p}(x_j) \mathbf{p}^T(x_j) \; ; \; \mathbf{B} = \left[\varphi(x_1) \mathbf{p}(x_1), \varphi(x_2) \mathbf{p}(x_2), \cdots, \varphi(x_n) \mathbf{p}(x_n) \right] \quad (15)$$

Matrix $\mathbf{A}$ may be written as

$$\mathbf{A} = [\mathbf{p}_1, \mathbf{p}_2 \cdots \mathbf{p}_n] \begin{pmatrix} \varphi(x_1) & 0 & \cdots & \cdots \\ 0 & \varphi(x_2) & 0 & \cdots \\ \cdots & \cdots & \ddots & \cdots \\ \cdots & \cdots & 0 & \varphi(x_n) \end{pmatrix} \begin{bmatrix} \mathbf{p}_1^T \\ \mathbf{p}_2^T \\ \vdots \\ \mathbf{p}_n^T \end{bmatrix} \qquad (16)$$

where

$$\mathbf{p}_j = \mathbf{p}(x_j) = \begin{Bmatrix} p_1(x_j) \\ p_2(x_j) \\ \vdots \\ p_m(x_j) \end{Bmatrix} \qquad (17)$$

The final approximation is still given by eq. (10) now however substituting matrix $\mathbf{C}$ by $\bar{\mathbf{C}}$. The new shape functions are therefore

$$N_j^i(x) = \sum_{l=1}^{m} p_l(x) \bar{C}_{lj}^{-1} = \mathbf{p}^T(x) \bar{\mathbf{C}}^{-1} \qquad (18)$$

It must be noted that accordingly to the least square character of the approximation

$$u(x_j) \simeq \hat{u}(x_j) \neq u_j^h \qquad (19)$$

i.e. the local values of the approximating function do not fit the nodal unknown values (Figure 1). Indeed $\hat{u}$ is the true approximation for which we shall seek the satisfaction of the differential equation and boundary conditions and u_j^h are simply the unknown parameters sought!

However if $n = m$ the FEM type approximation is recovered. Then $\hat{u}(x_j) = u_j^h$ and once again conditions (12) are satisfied.

2.1 Fixed Least Square Approximation (FLS)

The weighted least square approximation described above, depends on a great extend on the shape and the way to apply the weighting function. The simplest way is to define a fixed function $\varphi(x)$ for each of the Ω_i interpolation domains (see Figure 1a).

Let $\varphi_i(x)$ be a weighting functions satisfying

$$\varphi_i(x_i) = 1 \quad ; \quad \varphi_i(x) \neq 0 x \in \Omega_i \quad ; \quad \varphi_i(x) = 0 x \notin \Omega_i \tag{20}$$

Then the minimization square distance becomes

$$J = \sum_{j=1}^{n} \varphi_i(x_j)(u_j^h - \hat{u}(x_j))^2 \quad \text{minimum} \tag{21}$$

The expression of matrices $\mathbf{A}$ and $\mathbf{B}$ in eq.(14) are now

$$\mathbf{A} = \sum_{j=1}^{n} \varphi_i(x_j)\mathbf{p}(x_j)\mathbf{p}^T(x_j) \quad ; \quad \mathbf{B} = [\varphi_i(x_1)\mathbf{p}(x_1), \varphi_i(x_2)\mathbf{p}(x_2)\cdots, \varphi_i(x_n)\mathbf{p}(x_n)]$$

$$\tag{22}$$

Note that according to (6), the approximate function $\hat{u}(x)$ is defined in each interpolation domain Ω_i. In fact, different interpolation domains can yield different shape functions N_j^i. As a consequence a point belonging to two or more overlapping interpolation domains has different values of the shape functions which means that $N_j^i \neq N_j^k$. The interpolation is now multivalued within Ω_i and, therefore for any useful approximation a decision must be taken limiting the choice to a single value [22].

Indeed, the approximate function $\hat{u}(x)$ will be typically used to provide the value of the unknown function $u(x)$ and its derivatives in only specific regions within each interpolation domain. For instance by using point collocation we may limit the validity of the interpolation to a single point x_i. It is precisely in this context where we have found this meshless method to be more useful for practical purposes.

2.2 Moving Least Square (MLS) Approximation

In the moving least square (MLS) approach the weighting function φ is defined in shape and size and is translated over the domain so that it takes the maximum value over the position identified by the coordinate x_k where the unknown function $\hat{u}$ is to be evaluated. Note that x_k is an arbitrary position and not necessary coincident with one of the x_i points defined in eq.(4).

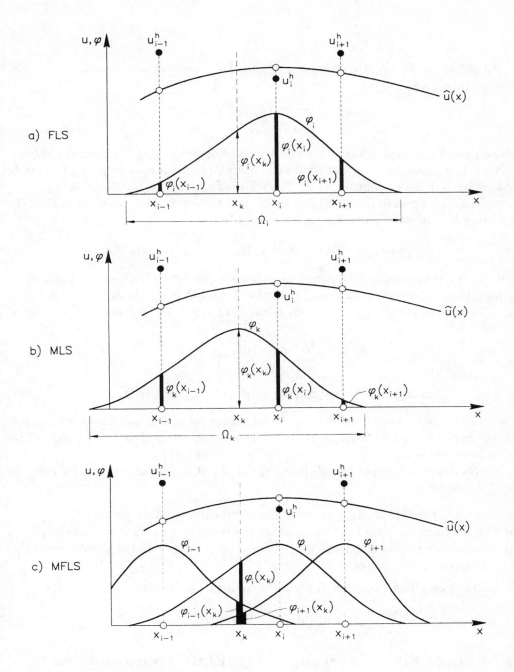

Figure 1. Different weighting least square procedures
a) Fixed least square (FLS); b) Moving least square (MLS);
c) Multiple fixed least square (MFLS)

As shown in Figure 1b we now minimize for any arbitrary coordinate x_k the following functional

$$J(x_k) = \sum_{j=1}^{n} \varphi_k(x_j)(u_j^h - \mathbf{p}_j^T \boldsymbol{\alpha})^2 \tag{23}$$

where φ_k can in general change its shape and span depending on the position of point x_k. Note again that x_k is now an arbitrary coordinate position and it can be simply replaced by the global coordinate x. We will however retain the form $\varphi_k(x_j - x_k)$ to emphazise the possibility of changing the function φ at each position within the approximation domain. Function φ_k is defined by

$$\varphi_k(x_k) = 1 \quad ; \quad \varphi_k(x) \neq 0 x \in \Omega_k \quad ; \quad \varphi_k(x) = 0 x \notin \Omega_k \tag{24}$$

where Ω_k is a subdomain of Ω around an arbitrary position x_k. Observe that now, J is a function of the position x_k, and then the $\mathbf{A}$ and $\mathbf{B}$ matrices are also a function of x_k. This must be taken into account when computing the derivatives of the shape functions, i.e.

$$\mathbf{A}(x_k) = \sum_{j=1}^{n} \varphi_k(x_j)\mathbf{p}(x_j)\mathbf{p}^T(x_j);$$

$$\mathbf{B}(x_k) = [\varphi_k(x_1)\mathbf{p}(x_1), \varphi_k(x_2)\mathbf{p}(x_2) \cdots, \varphi_k(x_n)\mathbf{p}(x_n)] \tag{25}$$

Note also that the parameters α_i are no longer constants but vary continuously with the position x_k and that inversion of matrices is required at every point where $\hat{u}$ is to be evaluated.

Furthermore, a unique global definition of the shape functions can be now obtained provided:

a) the weighting function φ_k is continuous and differentiable in Ω_k,
b) the weighting function φ_k vanishes on the boundary of Ω_k and outside,
c) the number of points n within Ω_k is equal or greater than the parameters m at all points in Ω_k. tex con142

In this case the MLS shape function of an arbitrary position x_k is

$$N_j^k(x) = \sum_{l=1}^{m} p_l(x)\bar{C}_{lj}^{-1}(x_k) \quad ; \quad \bar{\mathbf{C}}^{-1}(x_k) = \mathbf{A}^{-1}(x_k)\mathbf{B}(x_k) \tag{26}$$

2.3 Multiple Fixed Least Square (MFLS) Approximation

In general, with an arbitrary definition of points the problem of specifiying φ_k at every position x_k is very difficult and presents an infinite number of possibilities. In order to avoid this difficulty but preserving the unique global definition of the shape functions obtained in the MLS, Oñate *et al.* have recently propose a new possibility [22]. The idea is to define first a weighting function φ_i at any x_i point as in the FLS approximation. These functions are subsequently used to weight the square distances at an arbitrary position x_k. For a constant grid spacing with an invariant shape of the weighting function φ_k, it is possible to write: $\varphi_k(x_j) = \varphi_j(x_k)$

Introducing this equality in (23) we have

$$J(x_k) = \sum_{j=1}^{n} \varphi_j(x_k)(u_j^h - \mathbf{p}_j^T \boldsymbol{\alpha})^2 \qquad (27)$$

The least square problem leads to the values of $\boldsymbol{\alpha}$ given by eq.(14) with
$$\mathbf{A}(x_k) = \sum_{j=1}^{n} \varphi_j(x_k)\mathbf{p}(x_j)\mathbf{p}^T(x_j);$$

$$\mathbf{B}(x_k) = [\varphi_1(x_k)\mathbf{p}(x_1), \varphi_2(x_k)\mathbf{p}(x_2), \cdots \varphi_n(x_k)\mathbf{p}(x_n)] \qquad (28)$$

This method will be termed multiple fixed least square (MFLS) approximation since it makes use of different fixed functions φ_j to weight the square distances in $J(x_k)$ (Figure 1c).

This approximation is coincident with the MLS method just for an invariant shape of the weighting functions (i.e. $\varphi_k = \varphi_l = \varphi$). This algorithm produces also solutions for $\boldsymbol{\alpha}$ which depend on the position x_k. Note that the definition of the shape function is still *unique* and coincides with eq.(26) providing the three conditions of continuity and differentiability described for the MLS method are fulfilled.

3 Derivation of the discretized equations

The selection of different weighting functions in the general weighted residual form of eq.(3) yields different sets of discretized equations. In order to preserve the mesh-free character of the method, the weighting domain must be defined independently of any mesh. All approximation methods of integral type (i.e. Galerkin, area collocation, etc.) necessitate the introduction of complex procedures for integration (i.e. background grid, etc. [10-12, 16-19]). Some of these procedures are reviewed in [20,22]. In this paper we shall therefore limit the choice to *point collocation* methods where we feel the advantages of meshless procedures are best realized.

Such point collocation has recently been used with success for solution of inviscid/viscous flows by Batina [24] who however limited his work to the use of LSQ method and linear approximation. We have found that considerable improvement can be gained using weighted least square procedures [20-23].

Point collocation implies making $W_i = \bar{W}_i = \bar{\bar{W}}_i = \delta_i$ in eq.(3) where δ_i is the Dirac delta. This gives the set of equations

$$[A(\hat{u})]_i - b_i = 0 \quad \text{in } \Omega \qquad (29)$$
$$[B(\hat{u})]_i - t_i = 0 \quad \text{in } \Gamma_t \qquad (30)$$
$$\hat{u}_i - u_p = 0 \quad \text{in } \Gamma_u \qquad (31)$$

Any of the previous shape functions may be used now to approximate $\hat{u}$ leading in all cases to the system of equations

101

$$\mathbf{Ku}^h = \mathbf{f} \tag{32}$$

with $K_{ij} = [A(N_j)]_i + B(N_j)]_i$ and where the symmetry of the "coefficient" matrix $\mathbf{K}$ is not generally achieved. Vector $\mathbf{u}^h$ contains the problem unknowns, u_i^h, and $\mathbf{f}$ is a vector containing the contributions from the force terms b and t and the prescribed values u_p.

Taking a particular set of nodes and shape functions, this method is coincident with the generalized Finite Difference Methods of the type described in [5,7,8]. However we feel that the approach proposed here offers more possibilities. Indeed any of the interpolation techniques described in Section 2 can be used.

4 Stabilizing the finite point method

For non self adjoint problems such as occur in fluid mechanics special treatment is needed to stabilize the numerical approximation [1]. As a typical example we shall outline here the special feature on the convection-diffusion equation given by

$$A(\phi) = c\frac{\partial \phi}{\partial t} + \mathbf{u}^T \boldsymbol{\nabla} \phi - \boldsymbol{\nabla}^T(k\boldsymbol{\nabla}\phi) - Q = 0 \quad \text{in} \quad \Omega \tag{33a}$$

$$B(\phi) = \mathbf{n}^T k\boldsymbol{\nabla}\phi + \bar{q}_n = 0 \quad \text{in} \quad \Gamma_t \tag{33b}$$

$$\phi - \phi_p = 0 \quad \text{in} \quad \Gamma_u \tag{33c}$$

with the initial condition $\phi = \phi_0(\mathbf{x})$ for $t = t_0$.

In (33) $\boldsymbol{\nabla}$ is the gradient operator, c and k are known physical parameters, $\mathbf{u}$ is the velocity vector, ϕ the unknown field and Q a source term. $\bar{q}_n$ and ϕ_p are known values of the flux and the unknown function at the boundaries Γ_t and Γ_u, respectively.

Among the techniques typically used to stabilize FD, FE and FV methods we can list upwind finite difference derivatives, anisotropic balancing diffusion, Petrov-Galerkin weighting functions and characteristic time integration [1]. An alternative stabilization procedure appropiate for FPM which reproduces the best features of some of these techniques is described next.

4.1 Residual stabilization technique

A simple stabilization method can be derived by writting the flow balance equations in a finite domain [25]. The process is described in [31]for the simple 1D convection-diffusion problem leading to the following stabilized form of the steady state conservation equation

$$A(\phi) = r - \frac{h}{2}\frac{\partial r}{\partial x} = 0 \tag{34}$$

where

$$r = -u\frac{\partial \phi}{\partial x} + \frac{\partial}{\partial x}\left(k\frac{\partial \phi}{\partial x}\right) + Q \tag{35}$$

in eq.(34) h is the length of the domain (hereonwards termed *characteristic lenght*) where flow balance is enforced. This length can be written as $h = 2\tau u$ where τ is the so called *intrinsic time scale* in the FE convective transport and fluid flow literature [27]. The transient form of eq.(34) is readily obtained as

$$A(\phi) = c\frac{\partial \phi}{\partial t} - r + \frac{h}{2}\frac{\partial r}{\partial x} = 0 \tag{36}$$

The extension to 2D problems is straightforward as shown in [25] and gives

$$\text{steady state:} \qquad A(\phi) = r - \frac{h}{2|\mathbf{u}|}\mathbf{u}^T\boldsymbol{\nabla} r = 0 \tag{37}$$

$$\text{transient:} \qquad A(\phi) = c\frac{\partial \phi}{\partial t} - r + \frac{h}{2|\mathbf{u}|}\mathbf{u}^T\boldsymbol{\nabla} r = 0 \tag{38}$$

where in general

$$r = -\mathbf{u}^T\boldsymbol{\nabla}\phi + \boldsymbol{\nabla}^T k\boldsymbol{\nabla}\phi + Q \tag{39}$$

The stabilization term in eqs.(37) and (38) can be expressed in terms of the intrinsic time simply making $h = 2\tau|\mathbf{u}|$.

The discretization in time of eq.(38) can be written using a backward integration scheme as

$$\Delta\phi = \Delta t \left[r - \frac{h}{2|\mathbf{u}|}\mathbf{u}^T\boldsymbol{\nabla} r \right]^n \tag{40}$$

where $\Delta\phi = \phi^{n+1} - \phi^n$ and $(\cdot)^n$ denotes values computed at time t_n.

Eq.(40) can be found to be *identical* to that obtained using a *characteristic approximation* as described in [28] if the distance h is expressed by $h = |\mathbf{u}|\Delta t$.

The expression of the balancing term arising in above equations can be simplified if a linear or quadratic approximation is used as the cubic derivatives of ϕ in the diffusive term are zero. A further simplification arises if the source term is constant and the balancing term in eqs.(37) and (38) is then simply given by

$$\frac{h}{2|\mathbf{u}|}\mathbf{u}^T(\mathbf{u}^T\nabla\phi) \tag{41}$$

Note that the balancing term may be interpreted now as an additional diffusion as typically occurs in the FE literature [1].

The value of the optimal characteristic length h can be found by using the same arguments of standard upwinding and Petrov-Galerkin FE procedures, i.e. by searching exact nodal values for the simple 1D problem with $Q = 0$. Application of this concept to the FPM gives [20,22]

$$h = \frac{\alpha h}{2} \qquad \text{for linear interpolations } (m = 2); \alpha = \coth|Pe| - \frac{1}{|Pe|} \tag{42}$$

with the Peclet number defined as $Pe = \frac{|\mathbf{u}|\bar{h}}{2k}$.

In eq.(42) $\bar{h}$ is the distance measured along the streamline between the end points for a particular cloud.

4.2 Treatment of boundary conditions

The essential boundary conditions are simply satisfied pointwise on points x_i placed on Γ_u as

$$\phi(x_i) = \phi_p(x_i) \qquad x_i \text{ on } \Gamma_u \tag{43}$$

Other techniques to impose the essential boundary conditions are discussed on [22].

The authors have found that the straight forward satisfaction of the Neumann boundary condition in the FPM via eq.(30) leads to unstable results. This problem can be overcome by using the same residual stabilization technique as described above, applied now to the Neumann boundary condition.

This can be achieved by rewriting eq.(33b) (for the 1D case) as

$$B(\phi) = k\frac{\partial \phi}{\partial x} + \bar{q}_n - \frac{h}{2}r = 0 \tag{44}$$

where h is the *characteristic distance* defined in previous section.

Eq.(44) can be derived from the flow balance condition in a boundary domain of length $h/2$ as described in the Appendix where the extension to 2D problems is also shown. For further details see [25].

The new stabilized governing differential equations in the domain and the Neumann boundary conditions are discretized in space using the FP approximation as previously described.

Oñate [25] has shown that the application of the standard Galerkin FE weighted residual approach to the stabilized eqs.(37) and (44) leads to a system of discretized equations analogous to those obtained with the so called Petrov-Galerkin and SUPG FE procedures [1,27]. Indeed for certain values of the characteristic distance h all stabilization methods typically used in FE computations can be reproduced [25].

4.3 Compressible fluid flow

The approach proposed in previous section will be generalized now to solve two dimensional compressible fluid mechanics problems governed by the Navier-Stokes equations written in conservation form as [1]

$$\mathbf{A}(\mathbf{v}) = \frac{\partial \mathbf{v}}{\partial t} + \frac{\partial \mathbf{f}_i}{\partial x_i} + \frac{\partial \mathbf{g}_i}{\partial x_i} + \mathbf{q} = \mathbf{0} \qquad i = 1, 2, 3 \tag{45}$$

Equation (45) has been previously solved using the FPM with a Lax-Wendroff type scheme and linear base interpolations [23]. The solution will now be attempted using an extension of the residual stabilization approach presented in previous section and a *quadratic FLS interpolation*.

Following the same arguments of flow balance over a finite region given in previous section the following stabilized form of eq.(45) can be written as [25]

$$\mathbf{A}(\mathbf{v}) = \frac{\partial \mathbf{v}}{\partial t} - \mathbf{r} + \frac{h}{2|\mathbf{u}|} u_k \frac{\partial \mathbf{r}}{\partial x_k} = 0 \tag{46}$$

with

$$\mathbf{r} = -\frac{\partial \mathbf{f}_i}{\partial x_i} - \frac{\partial \mathbf{g}_i}{\partial x_i} - \mathbf{q} \tag{47}$$

where h is again the characteristic distance of the flow problem. A generalization of eq.(46) accounting for different values of h for each equation is proposed in [25]. A simple backward integration scheme leads to

$$\Delta \mathbf{v} = \Delta t \left[\mathbf{r} - \frac{h}{2|\mathbf{u}|} u_i \frac{\partial \mathbf{r}}{\partial x_i} \right]^n \tag{48}$$

where $\Delta \mathbf{v} = \mathbf{v}^{n+1} - \mathbf{v}^n$. Eq.(48) can be also derived using a characteristic approximation following the ideas described in [25]. A simpler form of eq.(48) neglecting the contribution of diffusive flows contribution in the stabilization term has been used by Zienkiewicz *et al.* [28,29] to derive a fractional step procedure to solve compressible and incompressible flows with the FEM.

Eq.(48) can now be discretized using the FPM by substituting the FP interpolation for the convective flux vector within each cloud as $\mathbf{f}_i = \sum_{j=1}^{n} N_j \mathbf{f}_i^j$. Choosing a point collocation procedure as described earlier leads to a system of equations from which the value of $\Delta \mathbf{v}$ (and subsequently those of $\mathbf{f}_i$ and $\mathbf{g}_i$) can be obtained and the solution advanced in time in the usual manner. Indeed the value of the time step increment Δt in eq.(48) must be adequately chosen to ensure stability of the time integration scheme [1,28,29]. Note that the treatment of the Neumann boundary condition requires again the use of the stabilization procedure described in previous section. This has however not been necessary for solution of the problem described next where only the essential boundary condition is imposed.

5 Concluding remarks

The paper shows that the fixed least square (FLS) interpolation combined with a point collocation tecnique is a promising Finite Point Method (FPM) for the solution of fluid mechanics problems. A residual stabilization technique which seems very adequate for the FPM has been proposed. It has been shown that in addition to the well known stabilization requirements for the governing differential equations over the internal domain, the Neumann boundary conditions need also to be stabilized. This has proved to be crucial to obtain acceptable results with the FPM even in cases where convection effects are neglected.

Excellent results have been obtained in all cases with quadratic base interpolations and clouds containing (at least) five and nine points for one and two dimensional problems, respectively.

References

[1] ZIENKIEWICZ, O.C. AND TAYLOR R.L. (1991) *The finite element method*, (Mc.Graw Hill, Vol. I., 1989, Vol. II

[2] ZIENKIEWICZ, O.C. AND OÑATE, E. (1991) Finite elements versus finite volumes. Is there a choice?", *Non Linear Computational Mechanics. State of the Art*, P. Wriggers and W. Wagner, eds., Springer-Verlag

[3] OÑATE, E., CERVERA, M. AND ZIENKIEWICZ, O.C. (1994) A finite volume format for structural mechanics, *Int. J. Num. Meth. Engng.*, **37**, 181-201

[4] IDELSOHN, S. and OÑATE, E. (1994) Finite element and finite volumes. Two good friends, *Int. J. Num. Meth. Engng.*, **37**, 3323-3341

[5] FORSYTHE, G.E. AND WASOW, W.R. *Finite-Difference Methods for Partial Differential Equations*, Wiley, New York

[6] PERRONE, N. AND KAO, R. (1975) A general finite difference method for arbitrary meshes, *Comp. Struct.*, **5**, 45-47

[7] LISZKA, T. AND ORKISZ, J. (1980) The finite difference method at arbitrary irregular grids and its application in applied mechanics, *Comp. Struct.*, **11**, 83-95

[8] LISZKA, T. (1984) An interpolation method for an irregular set of nodes, *Int. J. Num. Meth. Engng.*, **20**, 1594–1612

[9] MORAGHAN, J.J. (1988) An introduction to SPH, *Comp. Phys. Comm.*, **48**, 89-96, (1988).

[10] NAYROLES, B., TOUZOT, G. AND VILLON, P. (1992) Generalizing the FEM: Diffuse approximation and diffuse elements, *Comput. Mechanics*, **10**, 307-18

[11] BELYTSCHKO, T. LU, Y. AND GU, L. (1994) Element free Galerkin methods, *Int. J. Num. Meth. Engng.*, **37**, 229-56

[12] LU, Y., BELYTSCHKO, T. AND GU, L. (1994) A new implementation of the element free Galerkin method, *Comput. Meth. Appl. Mech. Eng.*, **113**, 397–414

[13] DUARTE, C.A. AND ODEN, J.T. (1995) H_p clouds-A meshless method to solve boundary-value problems, TICAM Repport 95-05

[14] BABUŠKA, I. AND MELENK, J.M. (1995) The partition of unity finite element method, Technical Note EN-1185, Inst. for Physical Science and Technology, Univ. Maryland

[15] TAYLOR, R.L., ZIENKIEWICZ, O.C., IDELSOHN, S. AND OÑATE, E. (1995) Moving least square approximations for solution of differential equations, *Research Report, 74*, International Center for Numerical Methods in Engineering (CIMNE), Barcelona

[16] LIU, W.K., JUN, S. AND ZHANG, Y.F. (1995) Reproducing Kernel particle methods, *Int. J. Num. Meth. Fluids*, **20**, 1081–1106

[17] LIU, W.K., JUN, S., LI, S., ADEE, J. AND BELYTSCHKO, T. (1195) Reproducing Kernel particle methods for structural dynamics, *Int. J. Num. Meth. Engng.*, **38**, 1655–1679

[18] LIU, W.K. AND CHEN, Y. (1995) Wavelet and multiple scale reproducing Kernel methods, *Int. J. Num. Meth. Fluids*, **21**, 901–933

[19] LIU, W.K., CHEN, Y., JUN, S., CHEN, J.S., BELYTSCHKO, T, PAN, C., URAS, R.A. AND CHANG, C.T. (1996) Overview and applications of the Reproducing Kernel particle methods, *Archives of Comput. Meth. in Engng.*, Vol. **3**, No. 1, 3–80

[20] OÑATE, E., IDELSOHN, S. AND ZIENKIEWICZ, O.C. (1995) Finite point methods in computational mechanics, *Research Report, 67,* (CIMNE, Barcelona

[21] OÑATE, E., IDELSOHN, S., ZIENKIEWICZ, O.C. AND FISHER, T. (1995) A finite point method for analysis of fluid flow problems, Proceedings of the 9th *Int. Conference on Finite Element Methods in Fluids*, Venize, Italy, 15-21

[22] OÑATE, E. IDELSOHN, S., ZIENKIEWICZ, O.C. AND TAYLOR, R.L. A finite point method in computational mechanics. Applications to convective transport and fluid flow, *Int. J. Num. Meth. Engng.*, **39**, pp. 3839-66, 1996

[23] FISHER, T., IDELSOHN, S. AND OÑATE, E. (1995) A meshless method for analysis of high speed flows, AGARD Meeting, Seville

[24] BATINA, J. (1993) A Gridless Euler/Navier-Stokes solution algorithm for complex aircraft applications, AIAA 93-0333, Reno NV, January 11-14

[25] OÑATE, E. (1996) On the stabilization of numerical solution of convective transport and fluid flow problems. *Research Report* No. 81, International Center for Numerical Methods in Engineering (CIMNE), Barcelona

[26] PERAIRE, J., PEIRO, J., FORMAGGIA, L., MORGAN, K. AND ZIENKIEWICZ, O.C. (1998) Finite element Euler computations in three dimensions, *Int. J. Num. Meth. Engng.*, **26**, 2135–59

[27] HUGHES, T.J.R. AND MALLET, M. (1986) A new finite element formulation for computational fluid dynamics. III: The generalized streamline operator for multidimensional advective-diffusive systems, *Comp. Meth. in Appl. Mech. and Engng.*, **58**, 305–28

[28] ZIENKIEWICZ, O.C. AND CODINA, R. (1995) A general algorithm for compressible and incompressible flow. Part I: The split characteristic based scheme, *Int. J. Num. Meth. in Fluids*, **20**, 869-85

[29] ZIENKIEWICZ, O.C., MORGAN, K., SATYA SAI, B.V.K., CODINA, RR. AND VAZQUEZ, M. (1995) A general algorithm for compressible and incompressible flow. Part II: Tests on the explicit form, *Int. J. Num. Meth. in Fluids*, **20**, No. 8-9, 886-913

[30] HIRSCH, C. (1990) *"Numerical computations of internal and external flow"*, Vol. 2, J. Wiley

[31] OÑATE, E., IDELSOHN, S., ZIENKIEWICZ, O.C., TAYLOR, R.L. AND SACCO, C. (1996) A stabilized finite point method for analysis of fluid mechanics problems, *Comput. Meth. Appl. Mech. Eng.*, **139**, 315–346

SYLVIE MAS-GALLIC

A vortex in cell method for the 2-d isentropic gas dynamic system

Abstract: The aim of this paper is to present a numerical method for the computation of the flow of a barotropic gas. We consider the system of equations constituted of the equation of conservation of mass and the equation of conservation of momentum. This system is written in velocity, density and pressure variable and the Helmholtz decomposition of the velocity together with some algebra yields to a system of equations written in vorticity, density and potential variables. This decomposition emphasises a natural separation between the quantities which are purely convected by the fluid and those which have an interaction with the fluid. The first ones are called passive whereas the second ones are named active quantities. The numerical method which is an extention of the *classical* P.I.C. method for incompressible fluids, is based on this separation : the density and the vorticity equations are solved by a lagrangian particle method in which the numerical particles carry both vorticity and density though the potential, the velocity and the *pseudo-pressure* are computed by means of *classical* eulerian methods.

Then, in the case of the modelisation of very small semiconductor devices, we present a model to which we apply the same decomposition method as previously and we are finally led to a system of equations very close the one constructed for the fluid model. It is thus natural to apply the numerical scheme constructed for the fluid model to the case of a semiconductor device.

1 Introduction

A lot of results are known about the flows of incompressible viscous fluids (see e.g. R. Temam [29]) or inviscid fluids (see C. Marchioro - M. Pulvirenti [20] and also G.H. Cottet [5], P.A. Raviart [28] for applications to particle methods) in several dimensions. In the case of compressible fluids, in dimension greater than one, only very few and partial results concerning the existence and regularity of flows of barotropic gas exist. Global existence of solution in the viscous case have been proved by P.L. Lions in [15] (see also [16] and [17] for related results). In the stationnary case still viscous, we also mention the work of A. Novotny [22] and A. Novotny - M. Padula [23]. The case of a slightly compressible flow can be found in C. Anderson [1] (see also A. J. Chorin [3]).

For an approach different from the one presented on a quite similar model we mention the work of L. Quartapelle [27] (see also more recently A. Ern [9]).

In the present paper, assuming existence of an exact solution we shall only be concerned with the presentation of a numerical method of resolution of the Euler system obtained from the mass and momentum conservation equations in the study of the flow of a barotropic gas. The method presented here is deeply inspired by the particle in cell method (P.I.C.) first introduced to compute flows of incompressible inviscid fluids and is now *classically* used (see F. Harlow [11], F. Harlow - A. Amsden [12], J.P. Christiansen [4] and more recently G.H. Cottet [6]). We recall that the P.I.C. method couples a lagrangian method, which solves the evolution equations of the physical quantities convected by the flow, and eulerian methods for the non transported quantities. In the work presented here, the philosophy of this very important idea was kept.

In the case of an incompressible fluid, written in velocity-vorticity variables $(\mathbf{u}, \omega)$, the Euler system reads

(1.1)
$$\begin{cases} \dfrac{\partial \omega}{\partial t} + \mathbf{u}.\nabla \omega = 0 \\ \nabla.\mathbf{u} = 0 \\ \nabla \times \mathbf{u} = \omega \end{cases}$$

now, we can rewrite the relation between ($\mathbf{u}$ and ω) as

$$-\Delta \mathbf{u} = \nabla \times \omega$$

and we notice that the vorticity is exactly and solely convected by the flow, that the pressure has disappeared and finally that the velocity is solution of an elliptic equation.

Thus, as soon as the system is written under this form, the numerical method consists first in discretizing the vorticity into vorticity elements which follow the characteristic curves of the velocity field and second in the approximation of the velocity on a grid so that the equation for $\mathbf{u}$ is solved by application of a standard finite difference scheme.

Let us mention that another formulation of the system expressed in current function - vorticity variables (ψ, ω), where the current function ψ is defined by $\mathbf{u} = \nabla \times \psi$ may seem more convenient especially for what concerns the boundary conditions (see the works of S. Huberson, A. Jollès and W. Shen [14]).

In the present case, we consider the flow of a two dimensional barotropic gas governed by the Euler system which results from the mass and momentum conservation. Denoting by ρ, $\mathbf{u}$ and p respectively the density, the velocity and the pressure of the gas and by γ its specific heat ratio, the equations of conservation of mass and of momentum as well as the barotropic gas law read

(1.2)
$$\begin{cases} \dfrac{\partial \rho}{\partial t} + \nabla.(\rho \mathbf{u}) = 0, \\ \dfrac{\partial \mathbf{u}}{\partial t} + (\mathbf{u}.\nabla)\mathbf{u} + \dfrac{1}{\rho}\nabla p = 0, \\ p = k\rho^{\gamma}. \end{cases}$$

(this system needs of course to be completed with initial conditions and boundary conditions that will be made precise later).

Since we have in mind to define an extension of the P.I.C. method to the calculation of the flow of such a fluid, we shall need first to separate (exactly as it was done in the incompressible case) between the quantities which are solely transported by the flow and those which interact with the flow. In order to attain this goal we shall now work on another formulation of the system. Using the Helmholtz decomposition of the velocity, we separate the velocity into a potential and a rotational part. Then, imbedding this decomposition into the equations, a little algebra will yield a system written in density, vorticity, velocity-potential, rotational velocity and *pseudo*-pressure variables.

From the resulting formulation, it appears clearly that the convected quantities are the density ρ and the vorticity ω whereas, the potential ϕ which satisfies a Hamilton-Jacobi type equation, as well as the rotational velocity $\mathbf{v}$ and the pseudo-pressure q which both are solution of elliptic equations, are not passively convected by the flow. Then, in order that the numerical coupled particle-grid method be well defined it is sufficient to make precise the schemes used. Let us mention that this formulation was first introduced in [19] (see also [21] and in the case of a potential flow [18]).

We consider from now on that the problem with periodic boundary conditions in the two directions. Since the method couples a particle method and a finite difference method, we shall need to define the two schemes used as well as the coupling operators which allow the exchange of information between the particles and the grid. The first operator which is the assignment operator of lagrangian values to the grid (vorticity and density for example) is due to J.U. Brackbill - H.M. Ruppel [2]. The choice of this assignment operator was based on its conservativity property but also on its easy extension to finite element methods. The second operator is a simple interpolation of Eulerian quantities (velocity for example) on the particles.

In the next Section, the formulation is presented precisely and the appearance of the variables is made clear. Section 3 is devoted to the presentation of the numerical scheme and some test results.

2 Vorticity-potential formulation of the system

Let us now present the system to be solved in its initial form. We start from the system (1.2) and consider the two dimensional case. As was said in the introduction, we are going to mimic the different steps of the PIC method when applied to the case of a compressible fluid and we introduce first the Helmholtz decomposition of the velocity field $\mathbf{u}$

$$(2.1) \qquad\qquad \mathbf{u} = \nabla\phi + \mathbf{v}, \qquad \nabla.\mathbf{v} = 0$$

where $\mathbf{v}$ is the rotationnal part of the velocity and ϕ the velocity potential. This decomposition separates the potential part, $\nabla\phi$, and the incompressible part, $\mathbf{v}$ of the velocity and by injecting it in system (1.2), we easily obtain the following system

(2.2)
$$\begin{cases} \dfrac{\partial\rho}{\partial t} + \nabla.(\rho(\mathbf{v}+\nabla\phi)) = 0, \\[2mm] \dfrac{\partial\mathbf{v}}{\partial t} - (\mathbf{v}+\nabla\phi)\times\nabla\times\mathbf{v} + \nabla q = 0 \\[2mm] \nabla.\mathbf{v} = 0 \\[2mm] \dfrac{\partial\phi}{\partial t} + \dfrac{1}{2}\mid \mathbf{v}+\nabla\phi\mid^2 + k\dfrac{\gamma}{\gamma-1}\rho^{\gamma-1} = q. \end{cases}$$

Thus, we take definitively advantage of the natural separation by introduction of the vorticity $\omega = \nabla\times\mathbf{u} = \nabla\times\mathbf{v}$ and we get the following system

(2.3)
$$\begin{cases} \dfrac{\partial\rho}{\partial t} + \nabla.(\rho(\nabla\phi+\mathbf{v})) = 0, \\[2mm] \dfrac{\partial\omega}{\partial t} + \nabla.(\omega(\nabla\phi+\mathbf{v})) = 0, \\[2mm] \dfrac{\partial\phi}{\partial t} + \dfrac{1}{2}|\mathbf{v}+\nabla\phi|^2 + k\dfrac{\gamma}{\gamma-1}\rho^{\gamma-1} = q, \\[2mm] \nabla.\mathbf{v} = 0, \\[2mm] \nabla\times\mathbf{v} = \omega, \\[2mm] \Delta q = \omega^2 + \nabla\omega\times(\nabla\phi+\mathbf{v}), \end{cases}$$

supplemented with initial conditions which read

$$\rho(.,0) = \rho_0, \quad \omega(.,0) = \omega_0, \quad \phi(.,0) = \phi_0.$$

After some algebra, we end up with the following new formulation of system (1.2)

(2.4)
$$\begin{cases} \dfrac{\partial\rho}{\partial t} + \nabla.(\rho(\nabla\phi+\mathbf{v})) = 0, & (i) \\[2mm] \dfrac{\partial\omega}{\partial t} + \nabla.(\omega(\nabla\phi+\mathbf{v})) = 0, & (ii) \\[2mm] \dfrac{\partial\phi}{\partial t} + \dfrac{1}{2}|\mathbf{v}+\nabla\phi|^2 + k\dfrac{\gamma}{\gamma-1}\rho^{\gamma-1} = q, & (iii) \\[2mm] -\Delta\mathbf{v} = \nabla\times\omega, & (iv) \\[2mm] \Delta q = \omega^2 + \nabla\omega\times(\nabla\phi+\mathbf{v}), & (v) \end{cases}$$

Now, under this form we notice that, as was mentionned in the introduction, both the density ρ and the vorticity ω are convected by the flow with the total velocity $\mathbf{u} = \nabla\phi + \mathbf{v}$. In the method that shall be presented later, the convection equations $(2.6)(i)$ and $(2.6)(ii)$ are solved by lagrangian methods whereas the equation $(2.6)(iii)$, $(2.6)(iv)$ and $(2.6)(v)$ are solved by finite difference schemes. Now, rewriting equation $(2.6)(iii)$ under the form

$$(2.5) \qquad \frac{\partial\phi}{\partial t} + \frac{1}{2}|\nabla\phi|^2 + \mathbf{v}.\nabla\phi = f,$$

where $f = q - 1/2|\mathbf{v}|^2$, we recognize a Hamilton-Jacobi equation with a convection term (see M.G. Crandall, P.L. Lions [7] for a detailed analysis). This equation will be numerically solved by a E.N.O. type scheme introduced by S. Osher and J. Sethian in [25] (see also [8], [26] and [27] for a stability analysis).

3 Numerical method

The proposed method is semi-lagrangian and needs thus the definition of both a finite difference grid and a set of weighted points (the particles).

First, given $\varepsilon > 0$, we fix a grid of width ε and denote by S_{ij} its vertices. The function ψ denotes the basis or interpolation function of the finite difference approximation (this function will be either piecewise constant in the case of N.G.P. method or piecewise bilinear in the case of C.I.C. method in the tests). At time 0 the discretized eulerian quantities, potential ϕ_h, rotational velocity $\mathbf{v}_h$ and pseudo-pressure q_h are respectively defined on the grid by

$$(3.1) \qquad \begin{cases} \phi_h^0(\mathbf{x}) = \sum_{ij} \phi_{ij}^0 \psi(\mathbf{x} - S_{ij}), \qquad q_h^0(\mathbf{x}) = \sum_{ij} q_{ij}^0 \psi(\mathbf{x} - S_{ij}) \\[2mm] \mathbf{v}_h^0(\mathbf{x}) = \sum_{ij} \mathbf{v}_{ij}^0 \psi(\mathbf{x} - S_{ij}). \end{cases}$$

We also need to define at time 0 approximations of the lagrangian quantities, the vorticity ω_h and the density ρ_h. Thus, we choose a set of weighted points $(W_p^0, \mathbf{x}_p^0)$ and define the following linear combinations of regularized Dirac measures as follows (also called particle approximation)

$$(3.2) \qquad \omega_h^0(\mathbf{x}) = \frac{1}{\varepsilon^2} \sum_p W_p^0 \omega_p^0 \psi(\mathbf{x} - \mathbf{x}_p^0), \qquad \rho_h^0(\mathbf{x}) = \frac{1}{\varepsilon^2} \sum_p W_p^0 \rho_p^0 \psi(\mathbf{x} - \mathbf{x}_p^0),$$

where the coefficients ω_p^0 and ρ_p^0 respectively stand for approximations of $\omega_0(\mathbf{x}_p^0)$ and $\rho_0(\mathbf{x}_p^0)$. Then, we introduce a time step $\Delta t > 0$, define a sequence of discrete times $t_n = n\Delta t$ and the discretization at time t_n of the equations of system (2.4).

As it was done at time 0, we define at time t_n, $n \geq 1$ the discrete eulerian quantities, ϕ_h^n, $\mathbf{v}_h^n$ and q_h^n by

$$
\begin{cases}
\phi_h^n(\mathbf{x}) = \sum_{ij} \phi_{ij}^n \psi(\mathbf{x} - S_{ij}), \qquad q_h^n(\mathbf{x}) = \sum_{ij} q_{ij}^n \psi(\mathbf{x} - S_{ij}) \\[2mm]
\mathbf{v}_h^n(\mathbf{x}) = \sum_{ij} \mathbf{v}_{ij}^n \psi(\mathbf{x} - S_{ij}).
\end{cases}
$$

The lagrangian quantities $\omega_h(.,t)$ and $\rho_h(.,t)$ are for any time $t > 0$ defined by a linear combinations of regularized Dirac measures as follows (also called particle approximation)

$$
\omega_h(\mathbf{x},t) = \frac{1}{\varepsilon^2} \sum_p W_p(t)\omega_p(t)\psi(\mathbf{x}-\mathbf{x}_p(t)), \quad \rho_h(\mathbf{x},t) = \frac{1}{\varepsilon^2} \sum_p W_p(t)\rho_p(t)\psi(\mathbf{x}-\mathbf{x}_p(t)),
$$

where $W_p(t)$, $\rho_p(t)$, $\omega_p(t)$ and $\mathbf{x}_p(t)$ are solutions of

(3.3)
$$
\begin{cases}
\dfrac{d\mathbf{x}_p}{dt}(t) = \mathbf{u}_h(\mathbf{x}_p(t),t) \\[3mm]
\dfrac{dW_p}{dt}(t) = \nabla.\mathbf{u}_h(\mathbf{x}_p(t),t)W_p(t), \\[3mm]
\dfrac{d\rho_p}{dt}(t) + \nabla.\mathbf{u}_h(\mathbf{x}_p(t),t)\rho_p(t) = 0. \\[3mm]
\dfrac{d\omega_p}{dt}(t) + \nabla.\mathbf{u}_h(\mathbf{x}_p(t),t)\omega_p(t) = 0.
\end{cases}
$$

(notice that $\omega_p(t)$ and $\rho_p(t)$ respectively stand for approximations of $\omega(\mathbf{x}_p(t),t)$ and $\rho(\mathbf{x}_p(t),t)$)

We now make precise the scheme at time t_n. First, the finite difference scheme for equation $(2.4)(iii)$ is the following extension of the ENO scheme (see S. Osher, J. Sethian [25])

(3.4)
$$
\frac{\phi_{ij}^{n+1} - \phi_{ij}^n}{\Delta t} = H_{H-J}(\bar{\phi})_{ij} + v_1^+ D_-^x \phi_{ij}^n + v_1^- D_+^x \phi_{ij}^n + v_2^+ D_-^y \phi_{ij}^n + v_2^- D_+^y \phi_{ij}^n + f_{ij}^n.
$$

where, setting $\bar{\phi}^n = (\phi_{ij}^n)$, the discrete derivative operators, Hamiltonian H_{H-J} and righthandside term are respectively defined by

(3.5)
$$
\begin{cases}
D_+^x \phi_{ij}^n = \dfrac{\phi_{i+1,j}^n - \phi_{ij}^n}{\Delta x}, \qquad D_-^x \phi_{ij}^n = \dfrac{\phi_{ij}^n - \phi_{i-1,j}^n}{\Delta x}. \\[3mm]
H_{H-J}(\bar{\phi})_{ij}^n = [|D_+^x \phi_{ij}^n|_+^2 + |D_-^x \phi_{ij}^n|_-^2] + [|D_+^y \phi_{ij}^n|_+^2 + |D_-^y \phi_{ij}^n|_-^2] \\[3mm]
f_{ij}^n = q_{ij}^n - k\dfrac{\gamma}{\gamma - 1}|\rho_{ij}^n|^{\gamma-1} - \dfrac{1}{2}|v_{ij}^n|^2.
\end{cases}
$$

This scheme is upwind though, as will be made precise in the next remark, in the potential case we define a centered scheme. This difference is mainly due to the appearance of the advection term in the Hamilton-Jacobi equation which needs the introduction of an extra treatment. This ENO scheme (as well as the scheme to be presented in the next remark) is stable under a CFL condition.

The two elliptic equations the solution of which are the rotational velocity $\mathbf{v}$ and the pseudo-pressure q are discretized either by the standard 5-points centered scheme or by a finite difference solver introduced by R. W. Hockney [13] (this last scheme is however specifically well adapted to the case of periodic boundary conditions).

It remains now to present the schemes for the time discretisation of the ordinary differential system of equations (3.3) and the *the particle-grid operator*. The density ρ_{ij}^n and vorticity ω_{ij}^n on the grid are defined by the assignment method due to Brackbill-Ruppel [2], for each grid point S_{ij}, we set

(3.6) $$\rho_{ij}^n = \frac{\sum_p W_p^0 \rho_p^0 \psi(S_{ij} - \mathbf{x}_p^n)}{\sum_p W_p^n \psi(S_{ij} - \mathbf{x}_p^n)}, \qquad \omega_{ij}^n = \frac{\sum_p W_p^0 \omega_p^0 \psi(S_{ij} - \mathbf{x}_p^n)}{\sum_p W_p^n \psi(S_{ij} - \mathbf{x}_p^n)}.$$

Now, an important remark needs to be done. First, since the fluid is compressible, we have seen in system (3.3) that the volume W_p of the particle located at point x_p evolve in time as well as the coefficient ρ_p and that their time evolution is given by the divergence of the velocity. However, their time evolution are opposite and it is clear from equations of system (3.3) that, their product $W_p\rho_p$ is constant in time and difference appears clearly in expression (3.5).

In order to complete the presentation of the scheme, let us make precise that the approximate positions of the particles as well as volumes are obtained by resolution of the system of ordinary differential equations appearing in system (3.3) by either a Runge-Kutta scheme of order 4 or the following Adams-Bashforth scheme

(3.7) $$\begin{cases} \mathbf{x}_p^{n+1} = \mathbf{x}_p^n + \Delta t(\frac{3}{2}\mathbf{u}_p^n - \frac{1}{2}\mathbf{u}_p^{n-1}) \\ W_p^{n+1} = W_p^n + \Delta t(\frac{3}{2}(\nabla_h.\mathbf{u}_h)_p^n W_p^n - \frac{1}{2}(\nabla_h.\mathbf{u}_h)_p^{n-1} W_p^{n-1}). \end{cases}$$

where $\mathbf{u}_p^n = \mathbf{v}_p^n + (\nabla_h\phi_h)_p^n$ and $\nabla_h\phi_h$ is the finite difference approximation of the gradient on the grid

Finally, the second exchange operator is a simple interpolation using either a piecewise constant function ψ (nearest grid point method, N.G.P.) or a piecewise bilinear interpolation function (cloud in cell method, C.I.C.) the results have been compared using one function or the other.

Remarks

    1. - In the case of a potential flow, the potential equation (2.6)(*iii*) were, is discretised by the classical leap frog scheme

$$\phi_{ij}^{n+1} = \phi_{ij}^{n-1} - \Delta t[(\frac{\phi_{i+1,j}^n - \phi_{i-1,j}^n}{2\Delta x})^2 + (\frac{\phi_{i,j+1}^n - \phi_{i,j-1}^n}{2\Delta y})^2 - 2k\frac{\gamma}{\gamma-1}|\rho_{ij}^n|^{\gamma-1}]$$

Notice that this scheme is now centered which is not the case of the scheme used in the general case.

    2. - It is also important to notice that the method of assignment defined in (3.5) is conservative (i.e. if $\rho_p \equiv 1$, then $\rho_{ij} \equiv 1$) which would not be the case if the following assignment method which is a lot simpler

$$\rho_{ij}^n = \sum_p W_p^0 \rho_p^0 \psi(S_{ij} - \mathbf{x}_p^n)$$

was used. From the physical point of view, this is specially dommaging when dealing with conservation equation.

    3. - Notice that the computation of $\nabla_h.\mathbf{u}_h$ in (3.6) needs the introduction of an extra scheme to compute $\Delta_h\phi_h$ and again the classical 5-points scheme is used.   ◻

    We present now some numerical results of two one dimensional tests, which represent namely, an infinite tube along the x-axis with two different sets of initial data. In the first test, illustrated by Fig.1, we start from a gas at rest with constant density ρ_0. This gas fills half of the tube and a piston limits it. At time 0, the piston is pulled on the left.

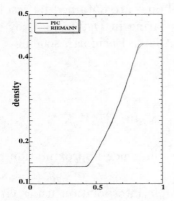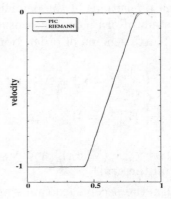

Fig.1: *Density and velocity profiles for PIC and Glaister's method [10] at*
$t = 1$ for $\Delta x = .01$ and $CFL = .1$

In the second test, illustrated by Fig.2, we start from a non uniform density which presents a shock.

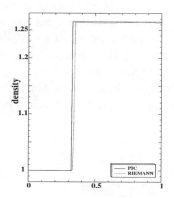 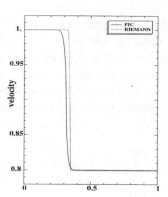

Fig.2: *Density and velocity profiles for PIC and Glaister's method at $t = 7.93$ for $\Delta x = .01$ and $CFL = .1$*

The results of these two tests show a good agreement with Glaister's solver although the PIC method is slightly more diffusive for the computation of both the density and the velocity in the first test and for the computation of the velocity in the second one. The reason for that may be that in the PIC method, the computation of the velocity was based on a Lax-Friedrichs scheme. Let us also point out that by contrast, the density-shock is better captured by the PIC method, illustrating the fact that particle methods are well suited for the resolution of transport equations.

4 Application to a semiconductor model

Let us briefly mention another possible application of this method in the context of very small semiconductor devices (Unterreiter [30]). Let us start from the one-particle, single state nonlinear Schrödinger-Poisson system

(4.1)
$$\begin{cases} i\hbar\dfrac{\partial\psi}{\partial t} + \dfrac{\hbar^2}{2m}\Delta\psi = eW\psi + k_B T \ln\left(|\psi|^2\right)\psi \\ -\varepsilon\Delta W = e(|\psi|^2 - C) \end{cases}$$

with unspecified boundary and initial conditions for the moment and let us make precise the different constant of the model : $\hbar$ is Planck's constant, m the electron mass, e the elementary charge and k_B Boltzmann's constant. The physical quantities that appear are : C, the doping profile, T, the electron temperature, ε, the electric permittivity of the device material.

117

This system of equations is submitted to several manipulations in order to see the analogy with the previous fluid model. First, rewriting the system in terms of particle density $n(x,t) = |\psi(x,t)|^2$ and particle current density $J(x,t) = \hbar/(2m)Im(\hbar\psi\nabla\psi)(x,t)$ we obtain, after some tedious algebraic manipulations, a system which, though equivalent to the initial nonlinear Schrödinger-Poisson model, does not contain any relaxation effect. Notice the importance of such effects for preventing /sl wild oscillations. Thus, the second manipulation actually consists in the addition of a relaxation term which leads to a new system no more equivalent to the initial model but which contains such searched effects. Finally, for the sake of a numerical study, we introduce the velocity field $\mathbf{u}$, $\mathbf{J} = n\mathbf{u}$ and the system becomes then

(4.2)
$$
\begin{cases}
\dfrac{\partial n}{\partial t} + \nabla.(n\mathbf{u}) = 0 \\[2mm]
\dfrac{\partial \mathbf{u}}{\partial t} + (\mathbf{u}.\nabla)\mathbf{u} + \dfrac{e}{m}\nabla W + \dfrac{k_B T}{m}\nabla \ln n - \dfrac{\hbar^2}{4m^2}\nabla(\dfrac{\Delta\sqrt{n}}{\sqrt{n}}) = -\dfrac{\mathbf{u}}{\tau} \\[2mm]
-\Delta W = \dfrac{e}{\varepsilon}(n - C).
\end{cases}
$$

Notice that this system has been obtained by division of the second equation by n which may include some problem whenever n vanishes. Now, arguing as in the case of the fluid equations of the previous section, we introduce the Helmholtz decomposition of the velocity $\mathbf{u}$ ($\mathbf{u} = \nabla S + \mathbf{v}$, $\nabla.\mathbf{v} = 0$), the vorticity $\omega = \text{curl } \mathbf{u} = \text{curl } \mathbf{v}$ and end up with the system

(4.3)
$$
\begin{cases}
\dfrac{\partial n}{\partial t} + \nabla.(n(\nabla S + \mathbf{v})) = 0 \\[2mm]
\dfrac{\partial \omega}{\partial t} + \nabla.(\omega(\nabla S + \mathbf{v})) + \dfrac{\omega}{\tau} = 0 \\[2mm]
\dfrac{\partial S}{\partial t} + \dfrac{1}{2}|\nabla S + \mathbf{v}|^2 + \dfrac{S}{\tau} + \dfrac{e}{m}W + \dfrac{k_B T}{m}\ln n - \dfrac{\hbar^2}{4m^2}(\dfrac{\Delta\sqrt{n}}{\sqrt{n}}) = q \\[2mm]
-\Delta \mathbf{v} = \text{curl } \omega \\[2mm]
-\Delta W = \dfrac{e}{\varepsilon}(n - C) \\[2mm]
-\Delta q = \omega^2 + \nabla\omega \times (\nabla S + \mathbf{v}).
\end{cases}
$$

Under this form we notice the obvious analogy of this system with system (2.4).

118

References

[1] C. ANDERSON, A Vortex Method for Flows with Slight Density Variations, J. Comp. Phys. **61**, 417-444 (1984).

[2] J.U. BRACKBILL, H.M. RUPPEL, Flip: A method for adaptively zoned, Particle-in-cell calculations of fluid flows in two dimensions, J. Comp. Phys. **65**, 314-343 (1986).

[3] A.J. CHORIN, Numerical Study of Slightly Viscous Flow, J. Fluid Mech. **57**, 785-796 (1973).

[4] J.P. CHRISTIANSEN, Numerical simulation of hydrodynamics by the method of point vortices, J. Comp. Phys. **13**, 363-379 (1973).

[5] G.-H. COTTET, Dynamique cochléaire en dimension un - Méthodes particulaires pour l'équation d'Euler dans le plan, Thèse de 3ème cycle, Paris (1982).

[6] G.-H. COTTET, Convergence of a Vortex In Cell method for the two-dimensional Euler equations, Math. Comp., **49**, 407-425, (1987).

[7] M.G. CRANDALL, P.L. LIONS, Math. Comp. **43**, 1 (1988).

[8] B. ENGQUIST, S. OSHER, Upwind difference schemes for systems of conservation laws-potential flow equations, M.R.C. technical report 2186 (1981).

[9] A. ERN, Sur la formulation tourbillon-vitesse des équations de Navier-Stokes avec densité et viscosité variables, C. R. Acad. Sci. Paris Série II (1997).

[10] P. GLAISTER, A Riemann solver for barotropic flow, J. Comp. Phys., **93**, 477-480, (1991).

[11] F. HARLOW, A. AMSDEN, Slip instability, Phys. of Fluids **7** , 327-334 (1964).

[12] F. HARLOW, The particle-in-cell computing method for fluid dynamics, Methods in Comp. Phys. **3** (1964).

[13] R.W. HOCKNEY, Comp. Phys. Comm. **2** (1971).

[14] S. HUBERSON, A. JOLLÈS, C. R. Acad. Sci. **309**, Série II, 445-448, Paris, 1989 and S. Huberson, A. Jollès and W. Shen, Numerical Simulation of Incompressible Viscous Flows by Means of Particle Methods in *Vortex Dynamics and Vortex Methods* (C. Anderson, C. Greengard ed.), A.M.S. Lectures in Applied Mathematics, vol **28**, 369-384, (1992).

[15] P.L. LIONS, Existence globale de solutions pour les équations de Navier-Stokes compressibles isentropiques, Note C. R. Acad. Sc. Paris (1993).

[16] P.L. LIONS, Compacité des solutions pour les équations de Navier-Stokes compressibles isentropiques, Note C. R. Acad. Sc. Paris (1993).

[17] P.L. LIONS, Limites incompressible et acoustique pour des fluides visqueux compressibles isentropiques, Note C. R. Acad. Sc. Paris (1993).

[18] M. LOUAKED, Une méthode P.I.C (particle-in-cell) pour les équations d'Euler compressibles, Rapport interne 90016, Lab. Anal. Num. (1990).

[19] M. LOUAKED, S. MAS-GALLIC, O. PIRONNEAU, A particle in cell method for the 2-D compressible Euler equations, in *Vortex Flows and Related Numerical Methods* (J.T. Beale, G.-H. Cottet and S. Huberson ed.), NATO ASI Series C, Mathematical and Physical Sciences **395**, Kluwer, Dordrecht (1993).

[20] C. MARCHIORO, M. PULVIRENTI, *Mathematical Theory of Incompressible Non-Viscous Fluids*, Applied Math. Sciences **96**, Springer New York (1993).

[21] S. MAS-GALLIC, A particle in cell method for the isentropic gas dynamic system, in *Navier-Stokes equations and related non linear problems*, (A. Sequeira ed.) New-York, Plenum Press (1995).

[22] A. NOVOTNY, Existence and uniqueness of stationary solutions for viscous compressible heat conductive fluid with large potential and small nonpotential forces, proc. EQUADIFF 91 (C. Perello, ed.), Barcelone (1991).

[23] A. NOVOTNY, M. PADULA, Existence and uniqueness of stationary solutions for viscous compressible heat conductive fluid with large potential and small nonpotential external forces, preprint 164, Univ. Ferrara (1991).

[24] S. OSHER, J. SETHIAN, Fronts propagating with curvature-dependent speed : Algorithms based on Hamilton-Jacobi formulations, J. Comp. Phys. **79**, 12-49 (1988).

[25] S. OSHER, F. SALOMON, Upwind difference schemes for hyperbolic systems of conservation laws, Math. Comp. **38**, (1982), 339-374.

[26] S. OSHER, C. SHU, High order essentially non-oscillatory schemes for Hamilton-Jacobi equations, Icase Report 90-13.

[27] L. QUARTAPELLE, Vorticity Conditioning in the Computation of Two-dimensional Viscous Flows, J. Comp. Phys. **40**, 453-477 (1981).

[28] P.-A. RAVIART, An analysis of particle methods, in *Numerical Methods in Fluid Dynamics* (F. Brezzi, ed.), Lecture Notes in Mathematics, vol. 1127, Springer Verlag, Berlin (1985).

[29] R. TEMAM, *Navier-Stokes equations*, North Holland, Amsterdam New-York, 77.

[30] A. UNTERREITER, (personnal communication)

MARIE POSTEL and MAURICIO SEPULVEDA

Modelling and numerical analysis for the propagation of a fluid in a porous medium

Abstract: Different reaction-diffusion macroscopic models of propagation of a fluid polymer in a chromatographic column are described along with numerical schemes. The link between the apparent diffusion and the microscopic-intrinsic diffusion of the fluid and the grains is recalled using the homogenization techniques for local equations. These intrinsic properties are sought after by parameter identification of the equation governing the global behavior. Numerical simulations are performed on experimental data, assuming periodic local geometry.

1 Introduction

The chromatography is a chemical separation process involving matter exchange between two distinct phases and used to analyse complex mixtures. To model it simply, we consider the propagation of one phase of a polymer in a porous medium, and we suppose the temperature to be constant. The equations of mass conservation are used with a nonlinear adsorption modelized by a function called *isotherm*: it represents a thermodynamical equilibrium state between the two phases.

Bear [1], Dagan [2], Gelhar and Axness [3] and Battacharya *et. al.* [4], take into account the microscopic structure of the porous medium and obtain relations between the effective – or macroscopic diffusion and the intrinsic – or microscopic one and also the speed of propagation. This approach is inadequate to explain the nonlinear behavior of the adsorption phenomenon. Thus, in a numerical point of view, we propose a more systematic study of the diffusion coefficients for the homogenized nonlinear equations modeling chromatography. The isothermal diphasic equilibrium is modeled by a relation $\omega_g = H(\omega_f)$ where ω_f is the concentration in the fluid phase, where ω_g is the concentration in the solid phase, and where the function H comes from statistical thermodynamics. The models of Langmuir [5], and Freundlich [6] are classically used in the literature. Taking the interaction between molecules into account, we use in this paper a generalization of the Langmuir isotherm (see [7]) :

$$
\begin{cases}
H(w) &= \dfrac{Kw \sum_{j=0}^{q-1} C_{q-1}^j b_{j+1}(Kw)^j}{\sum_{j=0}^{q} C_q^j b_j (Kw)^j} \\[2ex]
C_q^j &= \dfrac{q!}{j!(q-j)!} \\[2ex]
b_j &= \exp(-E_j/(RT))
\end{cases}
\tag{1}
$$

where K represents the Langmuir coefficient, and $\{E_p\}$, $p = 0, ..q$ the energetic coefficients (with $E_0 = E_1 = 0$). R and T are thermodynamical constants. Assuming the solid structure of the porous medium to be made of periodically distributed

121

symmetric grains, we can consider a one-dimensional global behavior, in which the fluid phase moves with a given velocity $u > 0$, and the solid phase is a stationary phase. We start with the work of Sepùlveda and James [8],[9] who model the propagation in a chromatography column of length L with a nonlinear hyperbolic equation. The nonlinear adsorption appears at the homogenized level on the left hand side of the equation:

$$\frac{\partial \omega(x, t)}{\partial t} + c\frac{\partial H(\omega(x, t))}{\partial t} = -\frac{\partial v(x)\omega(x, t)}{\partial x} \tag{2}$$

where $\omega(x, t)$ for $0 < x < L$ and $t > 0$ represents the concentration in the fluid phase. With this model a parameter identification is performed on the coefficients defining the isotherm function – in this case the Langmuir and the energetic coefficients. The coefficient c depends on the molecular structure and on the spatial concentration of grain in the column.

The hyperbolic model is realistic only when the diffusion effects can be neglected. This is not always the case in Chromatography experiments, and for instance in the real measurements we deal with in the later section. A first approach to consider the apparent diffusion in the measurements is to model it by the numerical viscosity of the numerical scheme (see [8]). It can be more or less precisely estimated - at least in the linear case - but it obviously depends on the choice of the discretization.

Another approach is to consider the same hyperbolic equation with an artificial viscosity term

$$\begin{cases} \dfrac{\partial \omega(x, t)}{\partial t} + c\dfrac{\partial H(\omega(x, t))}{\partial t} &= D\dfrac{\partial^2 \omega(x, t)}{\partial x^2} - \dfrac{\partial v(x)\omega(x, t)}{\partial x} \\ \omega(0, t) &= \omega_{in}(t) \\ \omega(x, 0) &= \omega_0(x) = 0. \\ \dfrac{\partial v(x)\omega(L, t)}{\partial x} &= 0 \end{cases} \tag{3}$$

Between (2) and (3) a diffusion term with a constant diffusion coefficient D has been added. If $D \to 0$, then it is proved that $\omega(D)$ converges to the solution of the hyperbolic equation (2) (see [10] and [11]). This convection-diffusion equation represents a model of chromatography with an infinite transfer velocity (see [12]).

In the numerical experiments, we will use this model on the same experimental measurements as in [8] and identify simultaneously the diffusion and the isotherm coefficients. We made a thorough study of the numerical behavior of (3) (see [13]). The convergence of the solution with respect to the space and time discretization is studied for a wide range of diffusion D and isotherm parameters. The choice of the discretization appears to be very important in the identification of the apparent diffusion, and also – if in a somewhat smaller respect – in that of the isotherm. More precisely, in [8], the discretization to solve the hyperbolic equation (2) is chosen so that the numerical viscosity best reproduces the diffusion apparent in the measurements. It turns out that for at least the dataset elicited for comparison of the two methods,

122

this discretization is too coarse to solve the convection-diffusion equation. In other words, using this discretization to identify the diffusion and the isotherm, will lead to parameters different from those using a finer discretization. The identified diffusion is smaller as it should because some of the phenomenon is accounted for by numerical viscosity. More important, the isotherm coefficients are also different from their limit values.

Having ascertained the need to model the apparent diffusion explicitely, it is now interesting to relate it to the microscopic structure of the porous medium. Vogt [14] obtains for a linear isotherm H the following homogenized equation

$$
\begin{cases}
\dfrac{\partial \omega(x,t)}{\partial t} + c\dfrac{\partial \eta(x,t)}{\partial t} &= D\dfrac{\partial^2 \omega(x,t)}{\partial x^2} - \dfrac{\partial v(x)\omega(x,t)}{\partial x} \\
\omega(0,t) &= \omega_{in}(t) \\
\omega(x,0) &= \omega_0(x) = 0. \\
\dfrac{\partial v(x)\omega(L,t)}{\partial x} &= 0
\end{cases}
\tag{4}
$$

where D is the homogenized diffusion coefficient and v is the homogenized speed obtained by standard homogenization of the Stokes problem in the interstitial domain. The function $\eta(x,t)$ acts on ω and takes into account the interaction between the grain and the fluid

$$
\eta(x,t) = \int_0^t \rho(t-\tau)\frac{\partial H(\omega(x,\tau))}{\partial t}d\tau
\tag{5}
$$

In the integral, H is the isotherm and $\rho(t)$ is the space average of the solution of a local problem on the micro-grain, which depends on the intrinsic microscopic diffusion in the grain D_g. Actually, the solution of (3) is the asymptotic behavior of the solution of (4) when $D_g \to \infty$ (see [11]). The apparent diffusion D is given by $D = D_f(1-\nu)$ where D_f is the intrinsic diffusion external to the grain, and ν is the mean of the solution of a Neumann cell problem characterized by the geometry of the grain (see [14] and [11]).

Remarks. When H is linear it is easy to generalize this homogenization result to a system of $2M$ transport equations and a vector of M concentrations representing a chemical mixture of M compounds. Canon and Jäger [15] prove the same result of homogenization in the scalar case for a nonlinear H using the so-called two scale convergence. The homogenization for a system and a nonlinear isotherm is an open problem.

Now, there is one more unknown to identify: the function of time ρ which acts as a sort of memory term and therefore enhances the intrinsic diffusion. As we will see in the simulations the apparent diffusion in equation (4) is due partly to the homogenized diffusion $D = D_f(1-\nu)$, partly to this convolution term. Our goal is to identify those two components. This is really the bottom line since we assume that the 'non-linear diffusion will be distinguishable from the classic one. Once ρ and D are satisfactorily

123

identified we are left with three unknowns: the intrinsic diffusions in the grain D_g and in the fluid D_f and the size of the grain R_g which must be identified using the two corresponding cell problems.

2   Numerical schemes

The stability and the convergence of possible schemes have been extensively tested in [13] for both problems (3) and (4). The spatial derivatives are treated implicitely and because of the high Peclet number value, up-wind discretization is used for the convection term (see [16]). For the convection-diffusion problem (3) the time derivatives are treated using backward finite differences over three time points as soon as all the input concentration has entered the column. During the injection a standard backward difference over two time points is used. Thus the resulting combined scheme preserves the conservativity of the time derivatives and convection term as in the initial equation. The nonlinear term is handled with a generalized Newton method.

The only originality in the numerical computation of (4) lies obviously in the treatment of η. Due to the convolution in time in (5) η acts as a memory term on ω. From the local problem defining ρ it is easily seen [11] that this function goes from zero at time zero to one as time tends to infinity. Furthermore, the more diffusion in the grain – the higher D_g is – the faster ρ nears its asymptotic value, one. In other words, the higher D_g is, the shorter is the memory. The function η is also nonlinear through the presence of the isotherm H in its definition. As it will be detailed in a later paragraph, ρ numerically reaches its asymptotic value quite fast. In other words, numerically the memory of the η process is finite and relatively short. From now on we will treat ρ as a given smooth function equal to 0 at the origin and equal to 1 after a finite time t_m. This simplification allows us to discretize the convolution as a finite sum of evaluations of H at the previous concentrations.

3   Parameter Identification

3.1   Delay function ρ

In the case of spherical grains the local problem giving the function $\tilde{\rho}$ can be exactly solved (see [14]). After averaging on the grain, the function ρ is

$$\rho(t) = 1 - \frac{2}{\pi} \sum_{n=1}^{\infty} \frac{1}{n^2} \exp(-D_g \pi^2 n^2 t) \qquad (6)$$

To use it numerically in the homogenized equation, we truncate it at a {sl convolution time } t_m after which it is taken constant equal to its asymptotic value 1. (In the following numerical experiments, the tolerance is chosen equal to 10^{-4}). Figure (1) shows on the left hand side the variation of t_m as a function of the intrinsic diffusion D_g.

On the right hand side the function $\rho(t)$ itself is represented for three different values of the intrinsic grain diffusion. For each curve the convolution time is indicated

124

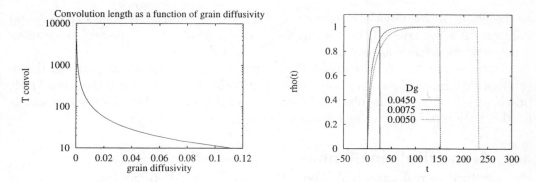

FIG. 1. Convolution length and function ρ for spherical grains

by a vertical line. In order to identify D_g, its physically possible range is systematically explored, starting from its limit infinite value which is treated using the convection-diffusion model (3). It is then decreased by varying increments corresponding to integer increments of the convolution time expressed in time-step units. For each guess value of D_g the identification of the other parameters of the model – the diffusion and the isotherm – is performed as explained in the next paragraph. Eventually the value of D_g leading to the minimum value of the identification criterium is the identified intrinsic grain diffusion.

3.2 Description of the identification algorithm

In a chromatography experiment, the concentration $\tilde{\omega}(L, t_i)$ is measured at the exit of the column $x = L$ at all time steps t_i for $i = 1, ...N$. We try to fit these curves with the solution $\omega(L, t_i)$ of either the homogenized equation (4) or the convection diffusion equation (3). The identification criterium is the L²-norm of the difference.

$$F(\{E_p\}, K, D, c) = \sum_{\text{exp.}} \left(\sum_{i=1}^{N} (\tilde{\omega}(L, t_i) - \omega(L, t_i)) \right)^2 \qquad (7)$$

The minimization of this functional can be done over all the parameters of the model, the diffusion D, the coefficients of the isotherm $K, \{E_p\}$ defined in (1) and the multiplying factor c.

First of all, a major difference with the identification scheme used in [8] is that we could not minimize (7) in a satisfactory way if we let the multiplying factor c set to its experimentally estimated value. The energetic coefficients thus identified are completely unphysical, trying to correct the 'bad guess made on c. This emphasizes a particular feature of the isotherm (1). One can easily check (see [13]) that numerically the energies do not define H uniquely in the range where it is used in practice. This range is actually very narrow (0 to 0.0006) for the experiment selected for the numerical application. On the other hand the energies really govern the asymptotic behavior

125

of the isotherm for large concentrations – for which unfortunately no experimental information is available. Thus it seems necessary to identify in a first stage the leading coefficients for small concentration – namely c and K.

A strategy which proves reasonable in practice is to use a Langmuir isotherm ($q=1$ in (1)) in a preliminary stage and identify D, K and c. Next the isotherm is progressively refined by increasing the degree q of the polynomial. The coefficient c is let this time equal to its value identified in the preliminary stage.

3.3 Numerical simulations

We now turn to real data, and choose the data set of n-Heptane (nC_7H_{16}) used in [8] for identification with the hyperbolic model. The best fit was there obtained using a 4th degree polynomial for the generalized isotherm (1). The discretization was $\delta_x = 7.10^{-4}m$, $\delta_t = 10$ seconds for an injection duration of 47 seconds and the identified parameters were $K = 1532$, $E2 = 825$, $E3 = 121$ and $E4 = 10$ with $c = 2.14$ being left fixed.

The preliminary identification described in the previous paragraph is performed using the convection-diffusion model and the homogenized one. For the latter, a discrete set of convolution lengths is explored – to which an intrinsic grain diffusion is associated as represented in Figure (1). We use the identified $K = 1532$ and $c = 2.14$ as initial guess for the identification with a time discretization $\delta_t = 5$ seconds, a space discretization $\delta_x = 0.06$ cm and a constant injection lasting 50 seconds. The second stage of the identification is then performed using for each trial value of intrinsic grain diffusion the identified value of the coefficient $c = 1.037$. The polynomial defining the isotherm is allowed to be of degree as high as 4-

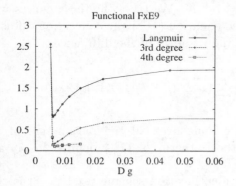
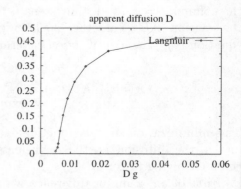

FIG. 2. Minimized functional and identified apparent diffusion as a function of intrinsic grain diffusion

On the left hand side of figure (2) the minimized functional F given by (7) is represented as a function of the intrinsic grain diffusion D_g in the homogenized model. The points $F(\infty) = 2.6E-9$ for the Langmuir case and $F(\infty) = 0.9E-9$ for the third degree lay outside the figure and are computed with the convection-diffusion model.

126

The identified parameters c, K, D are different for varying grain diffusion. The most obvious dependence is the one of the classical apparent diffusion. Its identified value using Langmuir isotherm is represented on the right hand side of figure (2). It decreases as the intrinsic grain diffusion decreases – which corresponds to an increasing convolution length – which makes sense since more dissipation is taken into account by the nonlinear adsorption phenomenon.

Figure (4) (resp. (3)) compares the experimental output concentrations with the values computed using the Langmuir isotherm (resp. the 4rd degree isotherm) and the identified parameters which in both cases correspond to an intrinsic grain diffusion of 0.0057. The concentrations computed by Sepùlveda using the hyperbolic model and the identified isotherm recalled above are also represented. The identified parameters are recalled in the following table.

Model	c	K	E2	E3	E4	D	D_g	F(x10^9)
Hyperbolic 4th degree	2.14	1532	825	121	10			
Homogenized Langmuir	1.037	3222	-	-	-	0.04	0.0059	0.85
Homogenized 3th degree	1.037	3152	-205	3000	-	0.0171	0.0059	0.16
Homogenized 4th degree	1.037	3114	-309	2814.	-784	0.0658	0.0065	0.10

The identified isotherms at the preliminary stage and with a polynomial of 4th degree are represented on figure (5). On the left hand side only the range of experimentaly available concentrations is represented. On the right hand side the isotherms are represented until they have reached their asymptotic value. It shows that even though the isotherms identified using the hyperbolic and the convection-diffusion models are globally very different, they are quite similar in the region of concentrations where they are used.

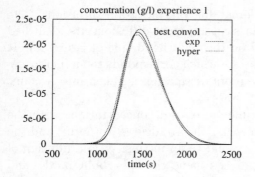

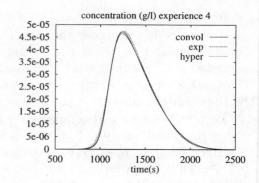

FIG. 3. Output concentrations – Experimental and computed with identified parameters of fourth degree isotherm

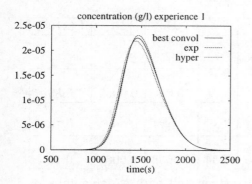

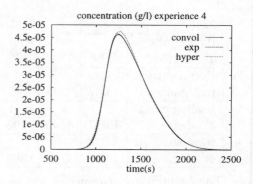

FIG. 4. Output concentrations – Experimental and computed with identified parameters of Langmuir isotherm

4 Conclusion

We proposed a numerical method to solve non-linear reaction-diffusion equations modeling the global behavior of a fluid in a chromatographic column. We used this scheme to identify the apparent diffusion coefficient and the parameters of the non-linear adsorption term, given an experimental data measured at the exit of the column. Furthermore, in the special case of spherical grains, the explicit relation between the nonlinearity in the homogenized equation and the intrinsic grain diffusion enables us to identify the latter.

For the kind of experimental data which we use here, the considered reaction-diffusion equations are more adapted than the hyperbolic model of chromatography. Moreover, we noted that the right modelization of the diffusion and the choice of a good discretization is important for the identification of the good parameters.

Here we have supposed spherical grains periodically distributed, which allows to reduce the homogenized problem to a one dimensional model. In general, it is known that the porous structure of chromatographic column is very heterogeneous and

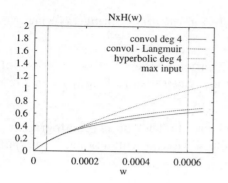 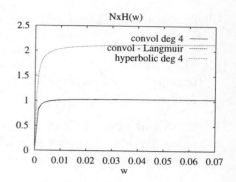

FIG. 5. Identified isotherms

intricate in its geometry. To deal with this, the full three dimensional homogenized equation must be considered and numerically solved. In fact the most realistic assumption of internal geometry could be to use a random model. This leads to other interesting developments both at the theoretical and numerical levels.

References

[1] J.BEAR, *Dynamics of Fluids in Porous Media*. American Elsevier, New York, 1972.

[2] G.DAGAN, Models of groundwater flow in statistically homogeneous porous formation. *Water Resources research*, 15(1):47–63, 1979.

[3] L.W.GELHAR AND C.L.AXNESS, Three-dimensional stochastic analysis of macrodispersion in aquifers. *Water Resources research*, 19(1):161–180, 1983.

[4] V.K.GUPTA, R.BHATTACHARYA AND H.F.WALKERS, Asymptotics of solute dispersion in periodic porous media. *SIAM J. Appl. Math*, 49(1):86–98, 1989.

[5] LANGMUIR I., *Jour. Am. Chem. Soc*, 38:2221–, 1916.

[6] FREUNDLICH, *Colloid and Capillary Chemistry*. Dutton, New York, 1926.

[7] P. VALENTIN, F. JAMES AND M. SEPÚLVEDA, Statistical thermodynamics models for a multicomponent two-phases equilibrium isotherm. *Maths. Models and Methods in Applied Science*, 1(7), 1997.

[8] M. SEPÚLVEDA, Identification de paramètres pour un système hyperbolique. application à l'estimation des isothermes en chromatographie. Thèse de doctorat, Ecole Polytechnique, France, 1993.

[9] M. SEPÚLVEDA AND F. JAMES, Parameters identification for a model of chromatography column. *Inverse Problems*, pages 1299–1314, 1994.

[10] J. SMOLLER, *Shock waves and recation diffusion equation.* Springer-Verlag, New-York, 1983.

[11] M. POSTEL AND M. SEPULVEDA, Identification of the effective diffusion for a transport equation modeling chromatography. *Submitted to Transport in Porous Media,* 1996.

[12] R. NABIL AND J. BARANGER, Sur quelques modèles de chromatographie en phase liquide. Rapport interne 90, Equipe d'Analyse Numérique, Lyon-Saint-Etienne, CNRS, URA 740, 1989.

[13] M. POSTEL AND M. SEPÚLVEDA, Numerical study of fluid propagation in chromatography column. parameter identification. Internal report, Universität Heidelberg- ELf Aquitaine, 1996.

[14] C.VOGT, A homogenization theorem leading to a volterra - integrodifferential equation for permeation chromatography. Preprint, Sonderforschungsbereich 123, Universität Heidelberg, FRG, 1982.

[15] W. JÄGER AND E. CANON, Homogenization of nonlinear adsorption-diffusion process in porous media. Preprint, Sonderforschungbereich 123, Universität Heidelberg FRG, Submitted to Asymptotic Analysis.

[16] S.V. PATANKAR, *Numerical heat transfer and fluid flow.* Series in Computational Methods in Mechanics and Thermal Science, Hemisphere Publishing Corporation, New-York, 1980.

MARIA-CECILIA RIVARA

Longest-edge algorithms for the refinement and/or improvement of triangulations

1 Introduction

In the adaptive finite element context, several mathematical algorithms for the refinement and/or derefinement of unstrutured triangulations, based on the bisection of triangles by its longest-edge, have been discussed and used in the last 10 years (Rivara 1984, 1989), (Rivara and Levin, 1992), (Rivara and Iribarren, 1996). These algorithms guarantee the construction of refined, nested and irregular triangulations of analogous quality as the input triangulation. However, the use of a new and related mathematical concept (the longest-edge propagation path of a triangle) recently introduced in (Rivara 1996a, 1996b), has allowed the development of new longest-edge algorithms for dealing with more general aspects of the mesh generation problem: 1) triangulation refinement problem, (2) triangulation improvement problem and (3) automatic quality triangulation of general geometries including small details. In this paper, different aspects of these mesh generation problems and some of the algorithms proposed to deal with them, are reviewed and discussed.

2 Mesh generation related problems

The polygon triangulation problem, an important issue in the use of finite element methods for engineering applications, can be formulated as follows:

DEFINITION 1. *Polygon Triangulation Problem: given N representative points of a polygonal region, join them by non intersecting straight line segments so that every region internal to the polygon is a triangle.*

Many criteria have been proposed as to what contitutes a"good" triangulation for numerical purposes, some of which involve maximizing the smallest angle or minimizing the total edge length. The Delaunay algorithm which constructs triangulations satisfying the first criteria has been of common use in engineering applications, followed by a postprocess step which assures the boundary respect of the polygon.

In the adaptive finite element context, the triangulation refinement problem has to be considered. To state this problem, some requirements and criteria about how to define the set of triangles to be refined and how to obtain the desired resolution need to be specified. To simplify we shall introduce a subregion R to define the refinement area, and a condition over the diameter (longest-edge) of the triangles (given by a resolution parameter ε) to fix the desired resolution.

DEFINITION 2. *Triangulation Refinement Problem: given an acceptable triangulation of a polygonal region Ω, construct a locally refined triangulation such that the* *diameters of the triangles that intersect the refinement area R are less than ε, and such that the smallest (or the largest) angle is bounded.*

In the particular case where the refinement is performed around any vertex, the point triangulation refinement problem can be stated.

DEFINITION 3. *Point Triangulation Refinement Problem: given an acceptable triangulation τ_0 of a polygonal region Ω, construct a (locally) refined triangulation τ, around any vertex P of τ_0, such that the diameter of the triangles of τ having the common vertex P are less than an input parameter ε.*

In the case we dispose of a bad-quality triangulation of the polygonal geometry (having a non-adequate distribution of vertices) the triangulation improvement problem has to be considered. To state this problem, a triangle quality indicator function q(t), a tolerance parameter ε, and a local triangle improvement criterion need to be specified. In this paper an LEPP point-insertion (improvement) criterion will be considered (see section 5).

DEFINITION 4. *Triangulation Improvement Problem: given a non-quality triangulation τ_0 of a polygonal region Ω (having triangles such that its quality indicator q(t) < ε), construct an improved triangulation τ such that each triangle t satisfies q(t) ≥ ε.*

Note that if an initial coarse triangulation of the boundary polygonal vertices is considered, the more general (automatic) quality polygon triangulation problem can be stated.

DEFINITION 5. *Quality Triangulation Problem: Given an initial (boundary) triangulation τ_0 of the boundary vertices which define the polygonal geometry, construct a geometry-adapted triangulation τ such that for each triangle t of τ, q(t) ≥ ε.*

At this point the following remarks are in order:

(1) The triangulation problems stated in Definitions 2 to 5 are essentially different than the classical triangulation problem in the following sense: insted of having a fixed set of points to be triangulated, one has the freedom to choose the points to be added in order to construct a mesh either with a desired resolution or with a given mesh-quality. The construction of the mesh is dynamically performed. Furthermore it is possible to exploit the existence of the reference triangulation (contructed for instance by means of the Delaunay algorithm) in order to reduce the computational cost to construct the output mesh.

(2) To cope with the triangulation Refinement Problem, the longest-edge refinement algorithms guarantee the construction of good quality irregular triangulations. This is due in part to their natural refinement propagation strategy farther than the (refinement) area of interest R. Asymptotically, the number N of points inserted in R to obtain triangles of prescribed size, is optimal. Furthermore, in spite of the unavoidable propagation outside the refinement region R, the time cost of the algorithm is linear in N, independent of the size of the triangulation. See section 6 for a review of some of these properties.

(3) In the remaining of this paper, new longest-edge based solutions both to the triangulation refinement problems of Definitions 2 and 3, and to the improvement and

quality triangulation problems of Definition 4 and 5 will be discussed. Note that in the latter cases, the algorithms take advantage of an LEPP point insertion technique (based on following the longest-edge propagation path of the target triangles) over Delaunay triangulations.

3   Longest-edge propagation path of a triangle

In this section we shall consider general conforming unstructured triangulations (where the intersection of adjacent triangles is either a common vertex or a common side). The following concepts have been implicitly used before in previous longest-edge refinement/derefinement algorithms (Rivara, 1984,1989):

DEFINITION 6. *For any triangle t_0 of any conforming triangulation τ, the Longest-Edge Propagation Path of t_0 will be the ordered list of all the triangles $t_0, t_1, t_2, ...t_n$, such that t_i is the neighbor triangle of t_{i-1} by the longest edge of t_{i-1}, for i=1,2,..., n. In addition we shall denote it as the LEPP(t_0).*

PROPOSITION 1. *For any triangle t_0 of any conforming triangulation of any bounded 2-dimensional geometry Ω, the following properties hold: (a) for any t, the LEPP(t) is always finite; (b) The triangles $t_0, t_1, ...t_{n-1}$ have strictly increasing longest side (if $n > 1$): (c) For the triangle t_n of the Longest-Edge Propagation Path of any triangle t_0, it holds that either: i) t_n has its longest edge along the boundary, and this is greater than the longest edge of t_{n-1}, or ii) t_n and t_{n-1} share the same common longest-edge.*

DEFINITION 7. *Two adjacent triangles (t,t*) will be called a pair of terminal triangles if they share their respective (common) longest edge. In addition, t will be a terminal boundary triangle if its longest-edge lies along a bundary edge.*

It should be pointed out here that the Longest-Edge Propagation Path of any triangle t corresponds to an associated polygon, which in certain sense measures the local quality of the current point distribution induced by t. To illustrates these ideas, see Figure 1, where the Longest-Edge Propagation Path of t_0 corresponds to the ordered list of triangles (t_0, t_1, t_2, t_3). Moreover the pair (t_2, t_3) is a pair of terminal triangles.

4   A backward longest-edge refinement algorithm

The original longest-edge refinement algorithms based on the longest edge bisection of triangles, were explicitly developed to solve the triangulation refinement problem in the adaptive finite element setting. The idea is to exploit the knowledge one has of the reference triangulation for working only locally with the refinement area (and some neighboring triangles). The new points introduced in the mesh are midpoints of the longest edge of (at least) one triangle of the reference mesh. In order to maintain a conforming triangulation, the local refinement of a given triangle involves refinement of the triangle itself and (longest-edge) refinement of some of its neighbors.

By using the LEPP(t) concept, a Backward Longest-Edge refinement algorithm

(Rivara 1996(a), 1996(b)) can be formulated, where the pure longest-edge refinement of a target triangle t_0 (see Figure 1) essentially means the repetitive longest-edge partition of the pair of terminal triangles associated with the current $\text{LEPP}(t_0)$, until the triangle t_0 itself is partitioned. Note that this backward algorithm indeed produces the same final triangulation as the original pure longest-edge algorithm, without intermediate non-conforming points in the mesh.

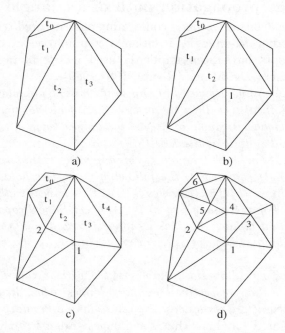

Figure 1. Backward Longest-Edge Bisection of triangle t_0
(a) Initial Triangulation. (b) First step of the process.
(c) Second step in the process. (d) Final triangulation

```
Backward-Longest-Edge-Bisection (t,T)
While t remains without being bisected do
   Find the LEPP(t)
   If t*, the last triangle of the LEPP(t), is a
   terminal boundary triangle, bisect t*
   Else bisect the (last) pair of terminal triangles of the LEPP(t)
```

The Figure 1 illustrates the refinement of the triangle t_0 over the initial triangulation of Figure 1(a) with associated $\text{LEPP}(t_0)=\{t_0, t_1, t_2, t_3\}$). The triangulations (b) and (c) illustrate the first 2 steps of the Backward-Longest-Edge-Bisection procedure and their respective current $\text{LEPP}(t_0)$, while that triangulation (d) is the final mesh obtained. Note that the new vertices have been enumerated in the order they were created.

134

The Backward-Longest-Edge-Bisection procedure is a non-recursive algorithm essentially based on refining pairs of terminal triangles (according to Definition 7). The concept of the Longest-Edge Propagation Path of the triangle t has been repeatedly used (over the current triangulation) in order to find the last 2 (terminal) triangles of the path, until the initial triangle i is bisected.

To solve the triangulation refinement problem, the algorithm guarantees the construction of good-quality irregular and nested triangulations, with linear time complexity, provided that an initial good quality triangulation is used (see section 6). To this end, a suitable data structure that explicitly manage the neighbor-triangle relation should be used. In addition, since at each iteration within the while loop, the LEPP(t) may or not be shortened, and may include new triangles not previously included in the LEPP(t) (see Figure1), the current LEPP(t) should be updated, rather than computed from scratch in order to get the linear running time. Furthermore, the new Backward Refinement Algorithm produces the same triangulation as the previous recursive algorithm, in a simpler, cleaner, easy-to-implement and more direct way.

5   Longest-edge improvement algorithm for Delaunay triangulations

The LEPP Delaunay improvement algorithm uses the Longest Edge Propagation Path of the target triangles (to be improved in the mesh) in order to decide which is the best point to be inserted to produce a good quality distribution of points (Rivara 1996a, 1996b). This algorithm generalizes the ideas introduced in (Rivara and Inostroza, 1995, 1997).

```
Basic Backward-LE-Delaunay-Improvement (t,T)
While t remains without being modified do
   Find the Longest Edge Propagation Path of t
   Perform a Delaunay insertion of the point p
   (midpoint of the longest edge of the last triangle
   in the LEPP(t))
```

Note that we have used the word improvement instead of bisection or refinement. This is to make explicit the fact that one step of the procedure does not necessarily produce a smaller triangle. More important however, is the fact that the procedure improves the triangle t in the sense of Theorem 6 of section 6.

For an illustration of the algorithm see Figure 2, where the triangulation (a) is this initial Delaunay triangulation with $LEPP(t_0) = \{t_0, t_1, t_2, t_3\}$, and the triangulations (b), (c) and (d) illustrate the complete sequence of point insertions needed to improve t_0. Note that in this example, the improvement (modification) of t_0 implies the automatic Delaunay insertion of 3 additional Steiner points. Each one of these points is the midpoint of the longest-edge of the last triangle of the current $LEPP(t_0)$. It should be pointed out here that each Delaunay point insertion locally improves the triangulation in the current LEPP(t_0), and in this sense this algorithm improves the triangulations obtained with the pure Backward Longest-Edge Refinement procedure.

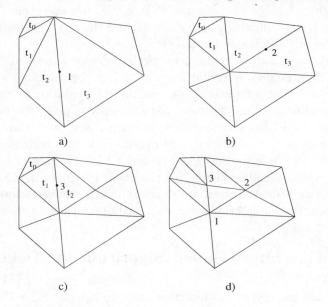

Figure 2. Backward Longest-Edge Delaunay improvement of triangle t_0

6   Mathematical properties of the backward techniques

6.1   The longest-edge refinement algorithms

The longest-edge refinement algorithms allow the local refinement of general (non-Delaunay and Delaunay) triangulations and produce irregular nested triangulations of analogous quality as the initial triangulation. The following Theorems summarize the properties of these algorithms, which are certainly valid for the backward version of the pure longest-edge algorithm described in section 4. For a detailed description of these results see references (Rivara 1984, 1987), (Rivara and Iribarren 1996), (Rivara and Venere, 1996).

THEOREM 1. *(a) The repetitive use of the pure longest-edge bisection algorithms, in order to refine t_0 and its descendants (triangles nested in t_0), tends to produce local quasi-equilateral triangulations. (b) The smallest angle α_t of any triangle t obtained throughout this process, satisfies that $\alpha_t \geq \alpha_0/2$, where α_0 is the smallest angle of t_0. (c) For any conforming triangulation τ, the global iterative application of the algorithm covers, in a monotonically increasing form, the area of t with quasi-equilateral triangles.*

Theorem 2 and 3 and Definition 8 summarize the fractal properties of the longest-edge refinement algorithms (Rivara and Venere, 1996). In particular, Theorem 3, assures that the longest-edge refinement algorithms always produce stable and

136

bounded "molecules" around each vertex of the triangulations (fixed partitions of the plane with bounded angles as illustrated in Figure 3).

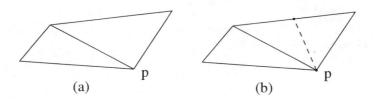

(a)   (b)

Figure 3 (a) Initial molecule associated with vertex P.
(b) Stable molecule associated with vertex P.

THEOREM 2. *Given any input triangle t of smallest angle α. For each vertex P of t (of interior angle δ), the arbitrary iterative use of the longest-edge refinement algorithm to solve the Point Triangulation Refinement Problem around vertex P, divides the angle δ exactly in k parts, where $k \leq 2\delta/\alpha_0$.*

DEFINITION 8. *For any conforming triangulation τ and any vertex P of τ, we shall call the stable molecule associated with vertex P to be the fixed partition of the plane around vertex P, induced by the use of the longest-side refinement algorithm, for solving the Point Triangulation Refinement Problem around P.*

THEOREM 3. *Let τ be any conforming triangulation and consider any vertex P of τ. The use of the longest-edge refinement algorithm to solve the Point Triangulation Refinement Problem around the vertex P, produces triangulations having the following characteristics:*

a) *After a finite number of iterations, the algorithm produces a triangulation τ^* such that, the stable molecule associated with vertex P is obtained.*

b) *The next iterations of the algorithm do not partition the angles of the stable molecule, but only introduce a set of new vertices distributed in geometric progression along the sides of the stable molecule of P.*

Theorem 3 allows to bound the number of vertices introduced in the Point Triangulation Refinement Problem (Theorem 4) and this result in turn can be used to prove that the time cost of the algorithms to solve the Triangulation Refinement Problem is linear (Theorem 5).

THEOREM 4. *Consider that the longest-edge refinement algorithm is used to solve the Point Triangulation Refinement Problem around any vertex P of any conforming triangulation τ. Then, once the stable molecule is obtained, the iterative refinement around vertex P, until obtaining triangles of diameter (longest side) h, introduces K vertices, where*

$K < k[1 + log_2(1 + 2L/h)],$

k is a constant and L is the length of the longest side of the stable molecule of P.

THEOREM 5. *Let τ_0 be any conforming coarse triangulation (of stable molecular size k) of any bounded polygonal region Ω. Then for any circular refinement subregion*

C, of radius r, with $C \subset \Omega$, the use of the longest-side refinement algorithm to solve the Triangulation Refinement Problem over C, until obtaining triangles of diameter $\tilde{h}$ inside C, asymptotically introduces N_i points inside C and N_o points ouside C, such that N_i and N_o satisfy that:

(a) $N_i = O(n^2)$

(b) $N_o = O(n \log_2 n)$

where $n = 2r/\tilde{h}$.

6.2 The LEPP Delaunay improvement algorithm

For the Backward LEPP Delaunay algorithm, I shall assume in addition that the input triangulation is a the constrained Delaunay triangulation of the input geometry data, which has a boundary point distribution that represents well the local feature size of the geometry boundary. This assumption allows to avoid boundary troubles which are surmounted in practice by using a boundary treatment technique described in (Rivara, 1996b).

In what follows I shall assume that t_0 is the worst triangle in the Longest-Edge Propagation Path of t_0. Note that this assumption is general enough, since if the LEPP(t_k) includes a worst triangle t_k, in this case we can first consider the smaller chain LEPP(t_k).

THEOREM 6. *For any Delaunay triangualation T, the repetitive use of the Backward-LE-Delaunay-Improvement algorithm over the worst triangles of the mesh (with smallest angle $\alpha < 30°$) produces a quality triangulation of smallest angles greater than or equal to $30°$).*

Proof. The proof is based on the properties of both the longest-edge refinement algorithms and the Delaunay triangulation. In effect, note that the pure Backward-Longest-Edge-Bisection procedure essentially adds to the current set of vertices, the midpoint of the longest-edge of the last greatest triangle of the LEPP(t), which in turn is inserted by longest-edge partition of the associated pair of terminal triangles of the current LEPP(t). This work produces nested triangles, and an adequate point distribution which guarantees that the percentage of good-quality triangles (and the area covered by these triangles) increases throughout the process (part(c) of Theorem 1). However, some bad triangles still remain in the mesh due to the fact that the longest-edge algorithms produce stable molecules around the vertices. In other words, as stated in Theorem 3, after a small number of partitions, the angles that share a vertex are fixed and not refined anymore. This is mainly due to the fact that nested triangulations are obtained.

When the Backward-LE-Delaunay-Improvement procedure is used in exchange, the midpoint of the longest-edge of the last greatest triangle of the LEPP(t) is also added, which improves the point distribution in the sense of the pure longest-edge algorithms. Futhermore, since this point is Delaunay inserted in the current triangulation, this local procedure improves the current triangulation in the following two senses: (1) the most equilateral mesh for the set of vertices is a obtained; (2) the worst angles of

138

the (non-fixed) molecules are eliminated (by edge swapping). If the triangle t is not destroyed thoughout the process, the new LEPP(t) is found over a locally improved triangulation; and as a consequence, the addition of the midpoint of the longest-edge of the last greatest triangle of the new LEPP(t), improves even more the current point distribution; while that the Delaunay insertion of this point again improves the triangulation in the two senses stated before; and son on. This process guarantees that the Backward-LE-Delaunay algorithm produces good-quality triangulations with smallest angles greater than or equal to 30°. Note that, smallest angles of more than 30° cannot be assured, since the longest-edge Delaunay partition of equilateral triangles can produce angles of 30° □

At this point the following remarks are in order:

(1) Even when Theorem 6 guarantees the construction of quality triangulations, it says nothing about the size of these triangualations. More mathematical results in this sense are certainly needed.

(2) In practice, the 2-dimensional triangulations obtained are size-optimal. They are of analogous quality as those obtained with the circumcenter point insertion strategy (Ruppert, 1995).

(3) The triangulation of Figure 4 illustrates the practical behavior of the algorithm. Note that the input data was the polygon with the minimun number of vertices to describe the geometry.

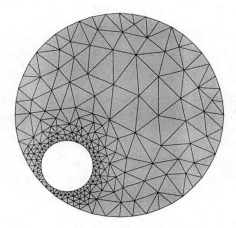

Figure 4 Automatic triangulation obtained
(smallest angles greater than 30°).

Aknowledgements. Some of the results presented in this paper were supported by FONDAP AN-1.

References

[1] RIVARA, M. C., Algorithms for refining triangular grid suitable for adaptive and multigrid techniques, International Journal for Numerical Methods in Engineering, 20(1984), 745-756.

[2] RIVARA, M. C., Selective refinement/derefinement algorithms for sequences of nested triangulations, International Journal for Numerical Methods in Engineering, 28(1989), 2889-2906.

[3] RIVARA, M. C., New mathematical tools and techniques for the refinement and/or improvement of unstructured triangulations, Proceedings 5th International Meshing Roundtable 96, Pittsburgh, 1996a, pp. 77-86.

[4] RIVARA, M. C., New longest-edge algorithms for the refinement and/or improvement of unstructured triangulations, International Journal for Numerical Methods in Engineering, 1996b, To appear.

[5] RIVARA, M. C. AND INOSTROZA, P., A discussion on mixed (longest side midpoint insertion) Delaunay techniques for the triangulation refinement problem, Proceedings 4th International Meshing Roundatable, Sandia National Labs, New Mexico, 1995, pp. 335-346.

[6] RIVARA, M. C. AND INOSTROZA, P., Using longest-side bisection techniques for the automatic refinement of Delaunay triangulations, International Journal for Numerical Methods in Engineering, 40(1997), 581-597.

[7] RIVARA, M. C. AND IRIBARREN, G., The 4-triangles longest-side partition of triangles and linear refinement algorithms, Mathematics of Computation, 65(1996), 1485-1502.

[8] RIVARA, M. C. AND LEVIN, C., A 3D refinement algorithm suitable for adaptive and multigrid techniques, Communications in Applied Numerical Methods, 8(1992), 281-290.

[9] RIVARA, M. C. AND VENERE, M., Cost analysis of the longest-side (triangle bisection) refinement algorithm for triangulations, Engineering with Computers, 12(1996), 224-234.

[10] RUPPERT, J., A Delaunay refinement algorithm for quality 2-dimensional mesh generation, Journal of Algorithms, 18(1995), 548-585.

MARKO A. ROJAS-MEDAR

On the existence of weak and strong solutions for the magneto-micropolar fluid equations in a time dependent domain

1 Introduction

In this paper, I report on joint work with J. L. Boldrini, C. Conca, P. Damásio, S. A. Lorca and E. Ortega-Torres. More details of our work can be found in [2], [23], [3], [8], [18], etc.

Many problems in fluid mechanics ocurr in time-varying regions. Such situations arise for instance in the case of

- A fluid in a vessel with moving boundaries.

- A fluid in a vessel containing rigid bodies moving through it.

Partial differential equations governing such phenomena are defined in a non-cylindrical domain. This leads to theoretical as well as numerical difficulties. In this conference we give a possible approach to respond the questions raised above. We show this with the magneto-micropolar fluid equations. It is important to observe that this technicality can be adapted for others equations of the fluid mechanic, such as, Boussinesq equations [2], [8], magnetohydrodynamics equations [21], viscous fluids in the pressence of difussion [3], nonhomogeneous incompressible fluids [9], [23], etc. The domain occupied by the fluid at time $t \in (0, T)$, $0 < T < \infty$, is denoted by $Q(t) \subseteq R^3$.

We set

$$Q = \bigcup_{0 < t < T} Q(t) \mathrm{x} \{t\} \subseteq I\!\!R^3 \mathrm{x} (0, T),$$

whose lateral boundary is

$$\partial Q = \bigcup_{0 < t < t} \partial Q(t) \mathrm{x} \{t\}.$$

Let $u(x, t) \in I\!\!R^3$, $w(x, t) \in I\!\!R^3$, $h(x, t) \in I\!\!R^3$ and $p(x, t) \in I\!\!R$, denote for $(x, t) \in Q$, respectively, the unknown velocity, the microrotational velocity, the magnetic field and the hydrostatic pressure of the fluid. Then, the governing equations are

$$\frac{\partial u}{\partial t} + u.\nabla u - (\mu + \chi)\triangle u + \nabla(p + \frac{1}{2}h.h) = \chi \operatorname{rot} w + r\, h.\nabla h + f,$$

$$j\frac{\partial w}{\partial t} + j\, u.\nabla w - \gamma \triangle w + 2\chi\, w - (\alpha + \beta)\nabla \operatorname{div} w = \chi \operatorname{rot} u + g, \qquad (1.1)$$

$$\frac{\partial h}{\partial t} + u.\nabla h - h.\nabla u - \nu \triangle h = 0,$$

$$\operatorname{div} u = \operatorname{div} h = 0,$$

141

together with the following boundary and initial conditions

$$u = 0, \ w = 0, \ h = 0 \text{ on } \partial Q, \tag{1.2}$$
$$u(x,0) = u_0(x), \ w(x,0) = w_0(x), \ h(x,0) = h_0(x), \ x \in \Omega(0). \tag{1.3}$$

Here μ, χ, r, α, β, γ, j and ν are constants associated to properties of the material. From physical reasons, these constants satisfy $\min\{\mu, \ \chi, \ r, \ j, \ \nu, \ \alpha+\beta+\gamma\} > 0$; $f(x,t)$ and $g(x,t) \in \mathbb{R}^3$ are given external fields.

For the derivation and physical discussion of equation (1.1)-(1.3) see Eringen [4], [5], Ahmadi and Shanhinpoor [1], for instance. Equations (1.1)(i) has the familiar form of the Navier-Stokes equations but it is coupled with equation (1.1)(ii), which essentially describes the motion inside the macrovolumes as they undergo microrotational effects represented by the microrotational velocity vector w. For fluids with no microstucture this parameter vasnished. For newtonian fluids, equation (1.1)(i) e (1.1)(ii) decouple since $\chi = 0$.

It is now appropiate to cite some earlier works on the initial-value problem (1.1)-(1.3) which are related to ours and also locate our contribution therein. In cylindrical domain and when the magnetic field is absent ($h \equiv 0$), the reduced problem was studied by Lukaszewicz [12], [13]. Lukaszewicz [12] stablished the global existence of weak solutions for (1.1)-(1.3) under certain assumptions by using linearization and an almost fixed point theorem. In the same case, by using the same technique, Lukaszewicz [13] also proved the local and global existence, as well as the uniqueness of strong solutions. Again when $h \equiv 0$, Galdi and Rionero [7] stablished results similar to the ones of Lukaszewicz [13].

The full system (1.1)-(1.3) in the cylindrical domain, was studied by Galdi and Rionero [7], and they stated without proofs results of existence and uniqueness of strong solutions. Rojas-Medar [19], Ortega-Torres and Rojas-Medar [17] and Rojas-Medar and Boldrini [22], also studied the system (1.1)-(1.3) and stablished the existence and uniqueness of local strong solutions, global strong solutions and existence of weak solutions, respectively, by using the spectral Galerkin method, reaching the same level of knowledge as in the case of the classic Navier-Stokes equations. Rojas-Medar [20] by using the results of [19] studied the convergences rates of spectral Galerkin approximations.

It has to be pointed out that similar time-dependent problems but for the Navier-Stokes equations have been studied by many different authors. This is the case, for instance, of the works by J. L. Lions [11] (see also the book of J. L. Lions [10]), H. Fujita and N. Sauer [6], H. Morimoto [27]. In particular, we would like to emphasize that the arguments in J. L. Lions [10] requires $Q(t)$ to be nondecreasing with respect to t (see problem 11.9, p. 426 of this book). Our paper, other than generalize these previous works in the sense that problem (1.1)-(1.3) includes the microrotational velocity and magnetic field, does not assume this nondecreasing condition on $Q(t)$.

This paper is organized in sections. After this brief introduction, in section 2, we introduce various function spaces. Next, section 3, we state the main theorems

142

of existence and/or uniqueness of solutions and finally in section 4, we give the error estimates of approximated solutions. The proofs of the results to appear in a next publication [18].

2 Function spaces and preliminares

The functions in this paper are either $I\!R$ or $I\!R^3$-valued and we will not distinguish these two situations in our notations. To which case we refer to will be clear from the context. Now, we give the precise definition of the time-dependent space domain $Q(t)$ where our initial boundary-value problems associated to the problem (1.1)-(1.3) has been formulated. Let $T > 0$, we consider the function $R : [0,T] \longrightarrow I\!R^9$, that is, $R(t)$ is a 3x3- matrix. Let Ω be an open bounded set of $I\!R^3$, which, without loss of generality, can be considered containing the origin of $I\!R^3$. We suppose that the boundary $\partial\Omega$ of Ω is smooth. We consider the sets

$$Q(t) = \{x = R(t)\,y \; ; y \in \Omega\,\}, \; 0 \le t \le T.$$

It is worth noting that such domains $Q(t), 0 \le t \le T$, generate a non-cylindrical time-dependent domain of $I\!R^3$ x $I\!R$

$$Q = \bigcup_{0<t<T} Q(t)\,\text{x}\,\{t\},$$

whose lateral boundary

$$\partial Q = \bigcup_{0<t<T} \partial Q(t)\,\text{x}\,\{t\}$$

is supposed regular. We make the following hypothesis on $R(t)$:

$$R(t) = r(t)\,M, \; \text{where} \; r : [0,T] \longrightarrow I\!R,$$

$r \in C^1([0,T])$, $r(t) > 0$, M is a 3x3 matrix whose entries are real constant and that there exist its inverse.

The main goal in this paper is to show existence of weak and strong solutions for the initial value problem (1.1)-(1.3). Our strategy for setting this question consists of transforming problem (1.1)-(1.3) into another initial-value problem in a cylindrical domain whose sections are not time-dependent. This is done by means of a suitable change of variable. Next, this new initial value problem is treated using Galerkin's approximation and the Aubin-Lions Lemma. We conclude returning to Q using the inverse of the above change of variable.

Sets of type (1) where $R(t) = r(t)\,I$, I identity 3x3-matrix, and Ω is the unit ball of $I\!R^3$ was considered by R. Del Passo and M. Ughi [26] to study a certain class of parabolic equations in noncylindrical domains.

Also, L. A. Medeiros and M. Milla-Miranda [14], [15] used the sets of type (1) where $R(t) = r(t)\,I$, and Ω is a bounded open set of $I\!R^n$, with regular boundary $\partial\Omega$

143

and $0 \in \Omega$ and $\min r(t) > 0$, to study exact contollability for Schröndinger equations in non-cylindrical domains.

C. Conca and Rojas-Medar [2] use the analogous domain that [26] to study the Boussinesq problem; M.A. Rojas-Medar and R. Beltrán-Barrios [21] for the magnetohydrodynamic type equations. The case of nonhomogeneous incompressible fluids was considered by J. Límaco-Ferrel [9], see also [23]. The formulation of the general class of domains considered in this paper was given by M. Milla-Miranda and J. Límaco-Ferrel [16] to study the classical Navier-Stokes equations . For other equations of fluids Mechanics to see [3], [8].

In order to state the main results we introduce some spaces; following the notation of [9], let $\mathcal{V}_t$ be the space

$$\mathcal{V}_t = \{\phi \in (C_0^\infty(Q(t)))^3 \, / \, \mathrm{div}\, \phi = 0\}$$

and $V_s(Q(t))$ be the closure of $\mathcal{V}_t$ in the space $(H^s(Q(t)))^3$, $s \in \mathbb{R}_+$. We use the particular notation

$$V_1(Q(t)) = V(Q(t)) \text{ and } V_0(Q(t)) = H(Q(t)).$$

The inner product of $V(Q(t))$ and $H(Q(t))$ are

$$((u,v))_t = \sum_{i,j} \int_{Q(t)} \frac{\partial u_i(x)}{\partial x_j} \frac{\partial v_i(x)}{\partial x_j} dx,$$

$$(u,v)_t = \sum_i \int_{Q(t)} u_i(x) v_i(x).$$

We observe that $V_s(Q(t)) \hookrightarrow (H_0^1(Q(t)))^3$ continuously for $s > \frac{1}{2}$ and

$$V(Q(t)) = \{u \in (H_0^1(Q(t)))^3 \, / \, \mathrm{div}\, u = 0\}.$$

We introduce in similar way the spaces $V_s(\Omega)$, in this case $\mathcal{V}$ has the form

$$\mathcal{V} = \psi \in (C_0^\infty(\Omega))^3 \, / \, \mathrm{div}\, (R^{-1}\psi^t) = 0\},$$

where ψ^t is the transposed of $\psi = (\psi_1, ..., \psi_3)$. We put

$$V_1(\Omega) = V \text{ and, } V_0(\Omega) = H$$

and

$$(u,v)_H = (u,v)_{L^2}, \quad (u,v)_V = ((u,v))_{L^2} = (\nabla u, \nabla v)_{L^2}.$$

In continuation, we will define the notion of weak and strong solutions for the problem (1.1)-(1.3).

144

Definition: We say that (u, w, h) is a **weak solution** of problem (1.1)-(1.3), if only if

$$u\,, h \in L^2(0, T; V(Q(t))) \cap L^\infty(0, T; H(Q(t))),$$
$$w \in L^2(0, T; H_0^1(Q(t))) \cap L^\infty(0, T; L^2(Q(t)))$$

and

$$(u_t, \varphi) + (\mu + \chi)\, a(u, \varphi) + b(u, u, \varphi) - rb(h, h, \varphi) = \langle f, \varphi \rangle + \chi(\mathrm{rot}\, w, \varphi), \quad (2.1)$$
$$j\,(w_t, \phi) + (Lw, \phi) + 2\chi(w, \phi) + j\, b(u, w, \phi) = \langle g, \phi \rangle + \chi(\mathrm{rot}\, u, \phi), \quad (2.2)$$
$$(h_t, \psi) + \nu\, a(h, \psi) + b(u, h, \psi) - b(h, u, \psi) = 0, \quad (2.3)$$

for all $\varphi, \psi \in V(Q(t))$ and $\phi \in H_0^1(Q(t))$ and

$$u(0) = u_0, \quad h(0) = h_0, \quad w(0) = w_0,$$

where

$$a(v, w) = \sum_{i,j=1}^{3} \int_{Q(t)} \frac{\partial v_j}{\partial x_i} \frac{\partial w_j}{\partial x_i} dx,$$

$$b(u, v, w) = \sum_{i,j=1}^{3} \int_{Q(t)} u_j \frac{\partial v_i}{\partial x_j} w_i dx,$$

$$Lw = -\gamma \Delta w - (\alpha + \beta) \nabla \mathrm{div}\, w.$$

Definition: Let $u_0, h_0 \in V(Q(0))$, $w_0 \in H_0^1(Q(0))$, $f, g \in L^2(Q)$. By a **strong solution** of problem (1.1)-(1.3), we mean functions u, w and h such that

$$u, h \in L^\infty(0, T; V(Q(t))) \cap L^2(0, T; H^2(Q(t)) \cap V(Q(t)))$$

$$\text{and } w \in L^\infty(0, T; H_0^1(Q(t))) \cap L^2(0, T; H^2(Q(t)) \cap H_0^1(Q(t)))$$

and that satisfies (2.1)-(2.3), for all $\varphi, \psi \in V(Q(t))$ and $\phi \in H_0^1(Q(t))$.

Our results are

Theorem 1. Under the above hypothesis on $Q(t)$ and if $f \in L^2(0, T; H(Q(t)))$, $g \in L^2(0, T; L^2(Q(t)))$, $u_0, h_0 \in H(Q(0))$ and $w_0 \in L^2(Q(0))$, then there exist a weak solution (u, w, h) of (1.1)-(1.3). Furthermore, $u, h \in L^\infty(0, T; H(Q(t)))$ and $w \in L^\infty(0, T; L^2(Q(t)))$.

Theorem 2. Let the initial values $u_0, h_0 \in V(Q(0))$, $w_0 \in H_0^1(Q(0))$ and the external forces $f, g \in L^2(0, T; L^2(Q(t)))$. Then, there exists $0 < T^* \le T$, such that problem (1.1)-(1.3) has a unique strong solution (u, w, h). This solution belongs to $C([0, T^*]; V(Q(t))) \mathrm{x} C([0, T^*]; H_0^1(Q(t))) \mathrm{x} C([0, T^*]; V(Q(t)))$.

Theorem 3. Let the initial values u_0, $h_0 \in V(Q(0)) \cap H^2(Q(0))$, $w_0 \in H_0^1(Q(0)) \cap H^2(Q(0))$ and the external forces f, g, f_t, $g_t \in L^2(0, T; L^2(Q(t)))$. Then the solution obtained in Theorem 2 satisfies

$$u, h \in C([0, T^*]; V(Q(t)) \cap H^2(Q(t))) \cap C^1([0, T^*]; H(Q(t))),$$

$$w \in C([0, T^*]; H_0^1(Q(t) \cap H^2(Q(t))) \cap C^1([0, T^*]; L^2(Q(t))).$$

Finally, the error estimates

Theorem 4. Under the hypothesis of Theorem 2, we have that there is a positive constant C, such that

$$\|u(t) - u^k(t)\|_{L^2(Q(t))}^2 \leq \frac{C}{\lambda_{k+1}}.$$

Here $u^k(t)$ is the Galerkin approximation of $u(t)$.

Acknowledgements. This research was supported by a grant from CNPq-Brazil, no. 351089/95-0.

References

[1] AHMADI, G. AND SHANHINPOOR, M., *Universal stability of magneto-micropolar fluid motions*, Int. J. Enging Sci., 12 (1974), 657-663.

[2] CONCA, C. AND ROJAS-MEDAR, M.A., *The initial value problem for the Boussinesq equation in a time-dependent domain*, Relatório Técnico de Pesquisa MA-93-B-402, Departamento de Ingeniería Matemática, Universidad de Chile, 1993. Submitted.

[3] DAMÁSIO, P. AND ROJAS-MEDAR, M.A., *The correctness of boundary-value problem in a diffusion model of an inhomogeneous liquid in non cylindrical domain*, preprint, 1995.

[4] ERINGEN, A.C., *Theory of micropolar fluids*, J. Math. Mech., 16 (1966), 1-8.

[5] ERINGEN, A.C., *Simple microfluids*, Int. J. Enging. Sci., 2 (1964), 205-217.

[6] FUJITA, H. AND SAUER, N., *Construction of weak solutions of the Navier-Stokes equation in a noncylindrical domain*, Trans. AMS 75(1969), 465-468.

[7] GALDI, G.P. AND RIONERO, S., *A note on the existence and uniqueness of solutions of the micropolar fluid equations*, Int. J. Enging. Sci, 15 (1977), 105-108.

146

[8] LORCA, S.A. AND ROJAS-MEDAR, M.A., *The initial value problem for a generalized Boussinesq model in a noncylindrical domain*, preprint, 1995.

[9] LÍMACO FERREL, J., *Existência de soluções fracas para a equação de fluidos viscosos incompressíveis não homogêneos em domínios não cilíndricos*, Tese de Doutorado, UFRJ-Brazil, 1993.

[10] LIONS, J.L., *Quelques méthodes de résolution des problèmes aux limites non linéaires*. Paris-Dunod, 1969.

[11] LIONS, J.L., *Une remarque sur les problèmes d'evolution non linéaires dans des domains non cylindriques*, Rev. Roumaine Math. Pures Appl., 9(1964), 11-18.

[12] LUKASZEWICZ, G., *On nonstationary flows of asymmetric fluids*, Rend. Accad. Naz. Sci. XL, Mem. Math., n. 106, vol. XII (1989), 83-97.

[13] LUKASZEWICZ, G., *On the existence, uniqueness and asymptotic properties of solutions of flows of asymmetric fluids*, Rend. Accad. Naz. Sci. XL, Mem. Math., n. 107, vol. XIII (1989),105-120.

[14] MEDEIROS, L.A. AND MILLA MIRANDA, M., *Exact controllability for Schöndringer equations in non cylindrical domains*, 41 Seminário Brasileiro de Análise (Minicurso), 1995.

[15] MILLA MIRANDA, M. AND MEDEIROS, L.A., *Contrôlabilité exacte l'equation de Schöndringer dans des domains non cylindriques*, CRAS-Paris, t. 319 Série I, 1994.

[16] MILLA MIRANDA, M. AND LÍMACO FERREL, J., *The Navier-Stokes in non-cylindrical domains*, 41 Seminário Brasileiro de Análise (1995).

[17] ORTEGA-TORRES, E.E. AND ROJAS-MEDAR, M.A., *Magneto-micropolar fluid motion: global existence of strong solutions*. Communication on this Congress.

[18] ORTEGA-TORRES, E.E. AND ROJAS-MEDAR, M.A., *On the initial value problem for the equations of magneto-micropolar fluid in a time-dependent domain*, R.P. 80/96, IMECC-UNICAMP, 1996.

[19] ROJAS-MEDAR, M.A., *Magneto-micropolar fluid motion: existence and uniqueness of strong solutions*, to appear in Mathematische Nachrichten.

[20] ROJAS-MEDAR, M.A., *Magneto-micropolar fluid motion: on the convergence rate of the spectral Galerkin approximations*, to appear in ZAMM.

[21] ROJAS-MEDAR, M.A. AND BELTRÁN-BARRIOS, R., *The initial value problem for the equations of magnetohydrodynamic type in non-cylindrical domains*. Revista de Matemática de la Universidad Complutense de Madrid, 8(1), 1995, 229-251.

[22] ROJAS-MEDAR, M.A. AND BOLDRINI, J.L., *Magneto-micropolar fluid motion: existence of weak solutions.* R.P. 42/95, IMECC–UNICAMP, 1995.

[23] ROJAS-MEDAR, M.A., AND CONCA C., *On the motion of nonhomogeneous fluids in region with moving boundaries*, preprint, 1992.

[24] SALVI, R., *On the existence of weak solutions of a non-linear mixed problem for the Navier-Stokes equations in a time dependent domain*, J. Fac. Sci. Univ. Tokyo, Sect IA, Math. 32(1985), 213-221.

[25] TEMAM, R., *Navier-Stokes equations*, revised ed., Amsterdam, North-Holland, 1979.

[26] DAL PASSO, R. AND UGHI, M., *Problème de Dirichlet pour une classe d´ equations paraboliques non linéaires dans des ouverts non cylindriques*, CRAS-Paris, Série I T-308(1989), 555-558.

[27] MORIMOTO, H., *On the existence of periodic weak solutions of the Navier-Stokes equations in regions with periodically moving boundaries*, J. Fac. Sci. Univ. Tokyo Sect. IA 18, 499-524.

MARIO STORTI, SERGIO IDELSOHN and NORBERTO NIGRO

(SU+C)PG: A Petrov-Galerkin formulation for advection-reaction-diffusion problems

Abstract: In this work we present a new method called (SU+C)PG to solve advection-reaction-diffusion scalar equations by the Finite Element Method (FEM). The SUPG (for *Streamline Upwind Petrov-Galerkin*) method is currently one of the most popular methods for advection-diffusion problems due to its inherent consistency and efficiency in avoiding the spurious oscillations obtained from the plain Galerkin method when there are discontinuities in the solution. Following this ideas, Tezduyar and Park treated the more general advection-reaction-diffusion problem and they developed a stabilizing term for advection-reaction problems without significant diffusive boundary layers. In this work an SUPG extension for all situations is performed, covering the whole plane represented by the Peclet number and the dimensionless reaction number. The scheme is based on the extension of the super-convergence feature through the inclusion of an additional perturbation function and a corresponding proportionality constant. Both proportionality constants (that one corresponding to the standard perturbation function from SUPG, and the new one introduced here) are selected in order to verify the "super-convergence" feature. i.e. exact nodal values are obtained for a restricted class of problems (uniform mesh, no source term, constant physical properties). It is also shown that the (SU+C)PG scheme verifies the Discrete Maximum Principle (DMP), that guarantees uniform convergence of the finite element solution. Moreover, it is shown that super-convergence is closely related to the DMP, motivating the interest in developing numerical schemes that extend the super-convergence feature to a broader class of problems.

1 Introduction

In this paper we focus on the numerical solution of the advective-reactive-diffusive equation using the finite element method. Here diffusion, advection and reaction refer to those terms in the governing equation involving second, first and zero order derivatives of the unknown variable. This kind of equation represents a simplified model for several industrial processes, for example the simulation of electrophoresis separation phenomena and the operation of a great number of chemical reactors. In these processes both the concentration and temperature play the role of the unknown variable.

Now, let us take the steady, linear advective-reactive-diffusive equation. As it is well known, the numerical solution of the above equation using Galerkin formulations exhibits global spurious oscillations in advection dominated problems, specially in the vicinity of discontinuities. Such drawback can be overcome by the popular SUPG method [1]. This method stabilizes the numerical scheme by adding a perturbation to the weight function producing an oscillation free solution. This perturbation is proportional to the gradient of the standard interpolation function, so therefore for linear constant-size elements it is an odd function about the center node. The amount of perturbation to be incorporated is calculated as a function of the dimensionless Peclet number. By adding a shock capturing term one can preclude the overshoot and undershoot in the neighborhood of the discontinuities [2].

On the other hand, another kind of problems exist in reaction dominated problems associated with the existence of local oscillations, also near discontinuities, even in the absence of advection terms. Similarly to the advection dominated problems, Tezduyar and Park [3] added another perturbation to the weight function. They choose it to be proportional to the gradient of the standard weight function, like the perturbation added for advection dominated problems, but with a different proportionality constant. The importance of the reaction term can be quantified by a dimensionless number, which we call the *reaction number r* formed by the reaction constant, the advection velocity and the element length. This scheme, which is called DRD, is designed to give the nodally exact solution for the homogeneous, one-dimensional analysis when the reaction number is much greater than the others and there are no diffusive boundary layers. One of the most remarkable critics of the above scheme is that, when the advection and the reaction terms are important, it is very difficult to know what amount of each perturbation function should be added. Another important criticism stems from symmetry considerations under coordinate inversion $x \rightarrow -x$. In the reactive-diffusive case (null advection), the equation is invariant under this symmetry operation and it is clear that the weight function should be symmetric. Actually, this is not the case of this scheme.

In this direction this paper tries to give an answer to the above questions. We present a Petrov-Galerkin formulation proposing two different perturbations to the weight function, one of them is similar to that of SUPG scheme and the other one is symmetric. For advection-diffusion problems, the scheme is reduced to the standard SUPG scheme. On the other hand, for reaction-diffusion problems only the symmetric perturbation subsists and the scheme is called CPG from *"Centered Petrov-Galerkin"*. In intermediate situations the scheme is a combination of the two, and then the acronym (SU+C)PG. The proportionality constant for each perturbation depends on the two dimensionless numbers, Peclet number Pe and the reaction number r. We find two expressions $\alpha(\mathrm{Pe}, r)$, $\gamma(\mathrm{Pe}, r)$ similarly as with the magic function in SUPG, where α, γ are the proportionality constants for both perturbation terms. With this kind of solution, we can solve in an optimal way not only the limit cases $\mathrm{Pe} \rightarrow \pm\infty$, $r \rightarrow \infty$, but also the whole Pe-r plane.

2 Optimal numerical scheme for the one-dimensional problem

Let us consider the following simplified, one-dimensional form of the advection-reaction-diffusion equation:

$$
\begin{aligned}
-k\phi'' + u\phi' + c\phi &= f, && 0 \le x \le 1, \\
\phi(x = 0) &= \phi_0, && \\
\phi(x = 1) &= \phi_1,
\end{aligned}
\qquad (1, 1 - 3).
$$

where $k > 0$ represents the physical diffusivity, u the transport velocity, ϕ the scalar unknown variable, $c \ge 0$ the reaction constant and f the source term. The weak formulation is:

$$
\int_\Omega (kw_i'\phi' + uw_i\phi' + cw_i\phi)\, d\Omega + \sum_{e=1}^{N} \int_{\Omega_e} p_i(-k\phi'' + u\phi' + c\phi)\, d\Omega =
$$

$$
\int_\Omega \tilde{w}_i f\, d\Omega, \qquad i = 1, \ldots, N - 1. \quad (2)
$$

Where w_i are the standard linear trial functions, $\tilde{w}_i$ the weight functions:

$$
\begin{aligned}
\tilde{w}_i &= w_i + p_i, \\
p_i &= \alpha h w_i' + \gamma P_{2i},
\end{aligned}
\qquad (3)
$$

p_i is the perturbation term and i, e are node and element indices, respectively. The first term is the well known SUPG perturbation term, whereas the second one is the new perturbation function which is intended to stabilize the reactive effects. With different expressions for the proportionality constants α and γ we obtain the (SU+C)PG method as well as those ones found in the literature:

$$
\begin{aligned}
\alpha &= 0, & \gamma &= 0, & &\text{Galerkin,} \\
\alpha &= \coth \mathrm{Pe} - 1/Pe, & \gamma &= 0, & &\text{SUPG,} \\
\alpha &= \alpha_{\mathrm{DRD}}(r/4\mathrm{Pe}), & \gamma &= 0, & &\text{DRD,} \\
\alpha &= \alpha_{(\mathrm{SU+C})\mathrm{PG}}(\mathrm{Pe}, r), & \gamma &= \gamma_{(\mathrm{SU+C})\mathrm{PG}}(\mathrm{Pe}, r), & &\text{(SU+C)PG,}
\end{aligned}
\qquad (4.1 - 4).
$$

where α_{DRD} is given by:

$$
\alpha_{\mathrm{DRD}}(x) = \tfrac{1}{2}\{-\coth(x) + x\,[1/\sinh(x)^2 + 4/6]\}/[1 - x\,\coth(x)]. \quad (5)
$$

The dimensionless parameters Pe and r, quantifying advection and reaction, respectively, with respect to the diffusive term are defined as:

$$
\mathrm{Pe} = \frac{uh}{2k},
$$

$$
r = \frac{ch^2}{k}. \qquad (6)
$$

$\alpha_{\text{(SU+C)PG}}(\text{Pe}, r)$ and $\gamma_{\text{(SU+C)PG}}(\text{Pe}, r)$ for the (SU+C)PG are obtained requiring that the scheme must be superconvergent for all r, Pe and after some algebra one arrives to the following system:

$$g_{j1} = [4\text{Pe}(1 - \cosh(\lambda_j h)) - r \sinh(\lambda_j h)],$$
$$g_{j2} = 2[\cosh(\lambda_j h)(rm - P_2(0)) + 2\text{Pe}\, a \sinh(\lambda_j h) +$$
$$(P_2(0) - mr + ar)], \tag{7}$$
$$f_j = -2[\cosh(\lambda_j h)(r/6 - 1) + \text{Pe} \sinh(\lambda_j h) + (1 + r/3)],$$
$$\lambda_j h = \text{Pe} + (-1)^{j-1}\sqrt{\text{Pe}^2 + r},$$

for $j = 1, 2$. We can obtain a super-convergent scheme by choosing an arbitrary function P_2 and computing α and γ from the preceding expression. However, for an arbitrary choice we will find, in general, that the proportionality constants have singularities for certain values of Pe and r, so that a design of the P_2 function is needed to avoid singularities. First, note that the previous expressions for α and γ only involve 3 geometrical parameters of the perturbation function P_2, namely: $P_2(0)$, a and m, which are, respectively, the value of the function at the center node, its semi-area, and one half its first moment. A detailed study [4] shows that there are no singularities if $P_2(0) = 0$ and P_2 is negative definite, so that $m, a < 0$. The lowest order polynomial (inside each element) that fulfills this requirements is:

$$P_2(\xi) = -\frac{1}{4}(1 - \xi^2), \tag{8}$$

where ξ is the coordinate in the master element $|\xi| \leq 1$. The geometrical parameters corresponding to this function are: $P_2(0) = 0$, $a = -1/6$ and $m = -1/12$.

3 The Discrete Maximum Principle

It is well known that the continuum problem satisfies a maximum principle, which can be put in the following terms: if $f(x) \leq 0$ for all x, then ϕ attains its maximum at the boundaries. The question is whether the discrete scheme inherits this feature, i.e. if, for any $f(x) \leq 0$ the numerical solution satisfies: $\phi_i \leq \max\{\bar{\phi}\}$ for all i, where $\bar{\phi}$ is the value of ϕ at the boundaries. It has been shown [5] that the satisfaction of the DMP implies uniform convergence of the finite element solution. The region of stability in the (Pe, r) plane has been assessed for each method and is shown in figure 1. The region of stability for Galerkin is a triangle, for pure advection the range of admissible Peclets is $|\text{Pe}| < 1$, whereas for pure reaction we must have $r < 6$. SUPG has a much broader range of stability. Needless to say, for pure advection all the Pe axis is in the stable zone, and for $|\text{Pe}| > \text{Pe}_{\text{crit}}$ it is stable for all r. However, for pure reaction the range of stable reaction numbers is the same as

Galerkin ($r < 6$). On the other hand the DRD method has an stable region smaller than Galerkin's. It is shown elsewhere [6] that the (SU+C)PG scheme is stable for all Pe and r.

4 Numerical results

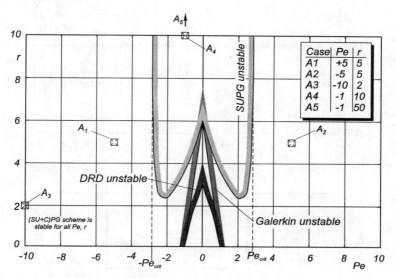

Figure 1: Stability map for Galerkin, DRD and SUPG.

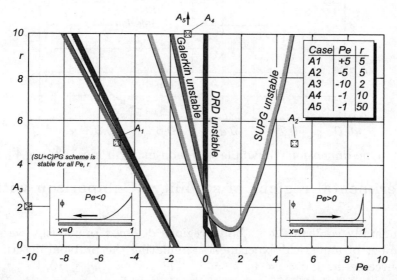

Figure 2: Experimentally determined stability map for Galerkin, DRD and SUPG.

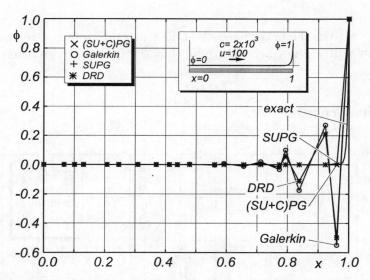

Figure 3: One-dimensional example with non-uniform mesh. Case A_1.

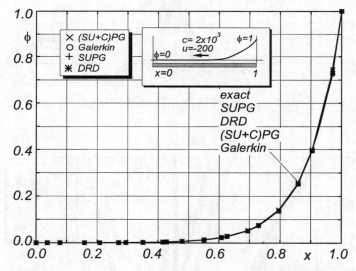

Figure 4: One-dimensional example with non-uniform mesh. Case A_2.

4.1 Experimental regions of stability. Comparison with other methods.

To experimentally confirm the stability regions deduced previously, we performed a large number of numerical experiments simulating the problem described by (1) with no source term, and $\phi_0 = 0$, $\phi_1 = 1$. The region $|\text{Pe}| \leq 10$, $0 \leq r \leq 10$ was covered with a grid of 100×50 points of the form $(\pm \text{Pe}_j, r_l)$ where the Pe_j and r_l, $j, l = 1, \ldots, 50$ are interpolated logarithmically between 0.2 and 10. The mesh was

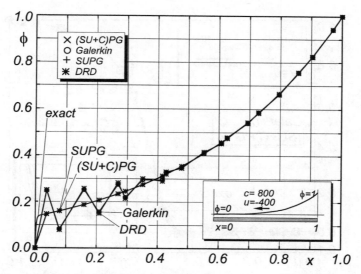

Figure 5: One-dimensional example with non-uniform mesh. Case A_3.

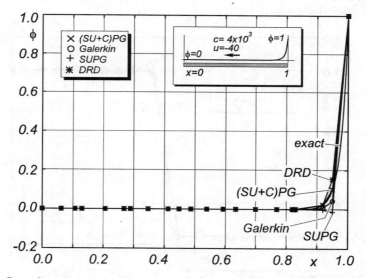

Figure 6: One-dimensional example with non-uniform mesh. Case A_4.

non-uniform, with 20 elements ($\bar{h} = \frac{1}{20} = 0.05$) and a maximun deviation with respect to the homogeneous $2|\Delta x_j|/\bar{h} < \delta_{\max} = 0.95$. The physical properties k, u and c were set according to the values of Peclet and reaction numbers, considered as based on the average size element $\bar{h}$. Since the exact solution, given by (9), is monotone we used this as the criterion for stability, i.e. the discrete solution $\{\phi_i\}$ for a given set of parameters (Pe_j, r_l) is unstable if $\phi_{i+1} < \phi_i$ for some i.

The resulting *stability map* is shown in figure 2. It can be seen that the right

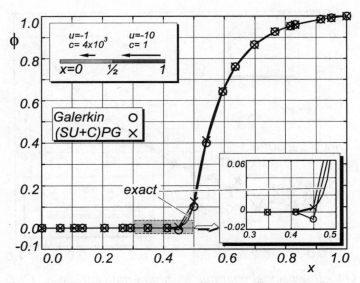

Figure 7: *Example with non-constant physical properties.Comparison between the different methods.*

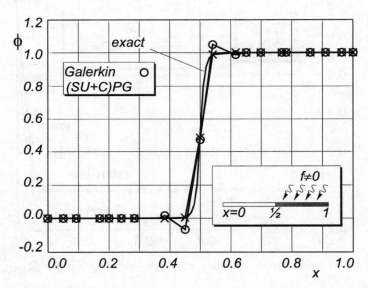

Figure 8: *Example with internal source and null advection. Comparison between the different methods.*

portion of the map (Pe > 0) is similar to the stability map based on the DMP (see figure 1). The differences can be attributed to the fact that the non-dimensional mesh parameters Pe and r are based in an average element size. In contrast, the left portion is qualitatively different from the DMP version. For all methods, the stability region for Pe < 0 is much broader than that for Pe > 0. This is because for

156

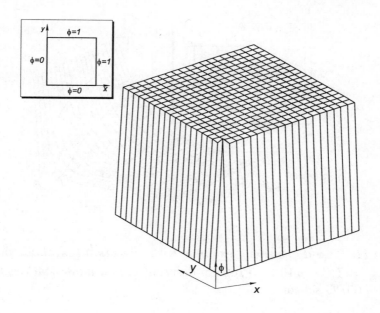

Figure 9: Two-dimensional example. Null advection, $r_h = 2.5 \times 10^5$. Stabilized scheme.

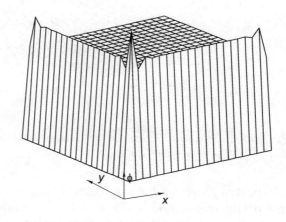

Figure 10: Two-dimensional example. Null advection, $r_h = 2.5 \times 10^5$. Galerkin.

a large part of the region Pe < 0, the solution is rather smooth, even if |Pe| and r are large. For instance, consider the cases Pe $= \pm 5$, $r = 5$ ($u = \pm 200$, $c = 2 \times 10^3$,

157

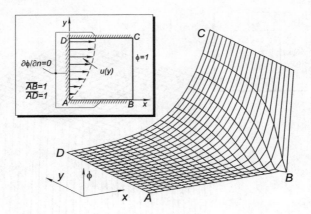

Figure 11: Two-dimensional example with advection (parabolic profile). $c = 5$, $u_{\max} = 1$, $k = 10^{-8}$. Problem description and numerical results with the (SU+C)PG scheme.

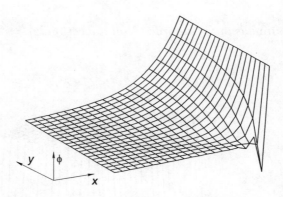

Figure 12: Two-dimensional example with advection (parabolic profile). $c = 5$, $u_{\max} = 1$, $k = 10^{-8}$. SUPG scheme.

as $\bar{h} = 1/20$), indicated as A_1 and A_2 in figure 2. The corresponding profiles can be seen in figures 3 (Pe $= +5$) and 4 (Pe $= -5$). The solution for Pe < 0 is rather smooth, since it is basically equal to the inviscid solution, which has a characteristic length of $\delta \approx |u|/c = 1/10$ and practically all methods give a qualitatively good result, even Galerkin. Actually, the DRD and SUPG method have a very small oscillation $\min_i\{\phi_{i+1} - \phi_i\} = O(10^{-6})$, produced by the boundary layer at the left boundary. Note that the position in the Pe-r plane is slightly inside the region of instability for both methods. In contrast, the solution for $u > 0$ exhibits a thin diffusive layer, with a characteristic length $\delta \approx k/|u| = 1/200$, and only the SUPG and

158

(SU+C)PG methods, which can cope with the strong diffusive boundary layer, give non-oscillatory results. Coming back to the Pe < 0 case, if we move deeper into the instability region of both the DRD and Galerkin methods, as for case A_3 (Pe = -10, $r = 2$, $u = -400$, $c = 800$, see figure 5)., then the amplitude of the mismatch at the left boundary is larger, and both DRD and Galerkin exhibit large oscillations. Up to this point, we have not found any advantage of DRD over SUPG, in fact DRD behaved very similarly to Galerkin. Consider now the case A_4 (Pe = -1, $r = 10$; $u = -40$, $c = 4 \times 10^3$, see figure 6).. The inviscid approximation continues to be valid and the downwind boundary layer is very small, but now the characteristic length for the inviscid solution is $|u|/c = 10^{-3} \ll \bar{h}$, so that the solution is not "smooth", and SUPG exhibits oscillations whereas DRD behaves very well. If we increase r to 50 ($c = 2 \times 10^4$) case A_5), the overshoot for SUPG is more pronounced, whereas DRD continues to be stable. This region, where $r \gg |Pe|$, and Pe < 0 is the region where DRD signifies an improvement over SUPG. However, as was shown, DRD will fail whenever a diffusive boundary layer is expected. In contrast, *the (SU+C)PG method was stable uniformly over the whole Pe-r plane.*

4.2 Non-constant physical properties

In this example, the physical properties u and c are not constant over the domain and non-uniform meshes are used: $N = 20$, $k = 1$, $\delta_{max} = 0.8$, $u = -1$ for $x < 0.5$ and $u = -10$; $x < 0.5$ $c = 4 \times 10^3$ for $x < 0.5$ and $c = 1$ for $x < 0.5$. The numerical solution is shown in figure 7, along with the Galerkin solution and the exact one. While the solution in the $x \geq \frac{1}{2}$ region is smooth, the solution in $x \leq \frac{1}{2}$ has a transition zone with a scale length 0.015, that is smaller than the average element size $\bar{h} = 0.05$. Consequently, Galerkin's method suffers from oscillations whereas the (SU+C)PG one remains stable.

4.3 Multi-dimensional results

The third test corresponds to a two dimensional example [7]. The uniform mesh is defined by 20 × 20 elements with a constant source term $f = 1$, null advection and $c = 1$, $k = 10^{-8}$, so that the element reaction number is equal to $r = 2.5 \times 10^5$. The boundary conditions are the following:

$$\phi(0, y) = \phi(x, 0) = 1 \qquad \text{for } 0 \leq x \, , \, y \leq 1,$$
$$\phi(1, y) = \phi(x, 1) = 0 \qquad \text{for } 0 \leq x \, , \, y \leq 1.$$

Figures 9,10 show the numerical results with the stabilized scheme and with the Galerkin scheme. The numerical oscillations disappear with the present method as in previous one-dimensional problems.

Finally we performed another two-dimensional example that consists of a linear convection reaction without source term but, now, the velocity is not constant [7]. The domain is the unit square $0 \leq x, y \leq 1$. The velocity $\mathbf{u} = (u, v)$ is assumed to have a parabolic profile:

$$u(y) = u_{\max}y^2, \quad v \equiv 0.$$

The reaction coefficient is equal to $c = 5$, diffusivity $k = 10^{-8}$ and $u_{\max} = 1$. The mesh is uniform and consists of 20×20 elements. The boundary conditions are natural (null flux) in three of the four sides of the domain and in the fourth side we have imposed a Dirichlet condition fixing the solution value equal to 1, as shown in figure 11. Figures 11, 12 show the stabilized scheme results and the standard SUPG ones. We can observe that the oscillations obtained with the standard SUPG near the zone where a very high reaction number exists have disappeared with the stabilized scheme whereas, in the rest of the domain, both solutions are equivalent since the problem is advection-dominated there.

5 Conclusions

This paper presents a stabilized multidimensional scheme for advection-reaction-diffusion problems that extends the SUPG method to overcome the numerical oscillations that appear from the reaction term. This numerical improvement allows us to treat, in an optimal way, a lot of industrial interesting situations where reaction effects are important in some regions of the domain, advection effects dominate in other zones, specially if the location of these zones is a priori unknown. It has been shown that (SU+C)PG scheme satisfies the DMP criterion uniformly over all the Pe-r plane and it has been proved stable also in a series of numerical tests.

Acknowledgments. The authors wish to express their gratitude to *Consejo Nacional de Investigaciones Científicas y Técnicas* (CONICET, Argentina) for its financial support. Software distributed from the *Free Software Foundation / GNU-Project* has been extensively used: Linux ELF-OS, Octave from John W. Eaton, Fortran f2c and g77 compilers, Tgif and others.

References

[1] A. BROOKS AND T. J. R. HUGHES, *Streamline upwind Petrov-Galerkin formulations for convection dominated flows with particular emphasis on the incompressible Navier-Stokes equations*, Computer Methods in Applied Mechanics and Engineering , vol.32, pp. 199-259 (1982).

[2] T. J. R. HUGHES AND M. MALLET, *A new finite element method for CFD: IV. A discontinuity-capturing operator for multidimensional advective-diffusive systems*, Computer Methods in Applied Mechanics and Engineering vol. 58, pp. 329-336, (1986).

[3] T. TEZDUYAR AND Y. PARK, *Discontinuity capturing finite element formulations for nonlinear convection-diffusion-reaction equations*, Computer Methods in Applied Mechanics and Engineering , vol.59, pp. 307-325 (1986).

[4] M. STORTI, N. NIGRO AND S. IDELSOHN, *A Petrov-Galerkin formulation for the advection-reaction-diffusion equation*, Revista Internacional de Métodos Numéricos para Cálculo y Diseño en Ingeniería, **11**, pp. 247-270, (1995)

[5] P.G. CIARLET AND P.A. RAVIART, *Maximum principle and uniform convergence for the finite element method*, Computer Methods in Applied Mechanics and Engineering , **2**, (1973), pp. 17-31

[6] S. IDELSOHN, N. NIGRO, M. STORTI AND G. BUSCAGLIA, *A Petrov-Galerkin formulation for advection-reaction-diffusion problems*, Computer Methods in Applied Mechanics and Engineering vol. 136, pp. 27-46 (96)

[7] R. CODINA, *A shock capturing anisotropic diffusion for the finite element solution of the diffusion-convection-reaction equation*, in Numerical Methods in Engineering and Applied Sciences, by E. Oñate (edt.), CIMNE, (Barcelona, Spain), (1993).

JEAN-MARIE THOMAS and DAVID TRUJILLO

Finite volume methods for elliptic problems: convergence on unstructured meshes

1 Introduction

The finite volume method is a discretization technique for conservation laws. The first proofs of convergence of finite volume methods for elliptic problems were given on regular structured meshes and founded on finite difference techniques (see Heinrich [9]). Unfortunately, the generalization to more complex geometry leads to some difficulties (see Gallouët [8] and Faille [7]).

Finite volume element methods were developed as an attempt at using finite element ideas to create a more systematic finite volume methodology. Thus, convergence results have been obtained for diffusion equations on triangular meshes with linear finite elements spaces (see Cai [4], Cai, Mandel & McCormick [5]). However, abstract results were established under a discrete uniformly elliptic assumption; this discrete assumption seems to be difficult to verify for a general self-adjoint elliptic boundary value problem discretized with unstructured meshes.

We give below the application of a new methodology called mixed finite volume method (MFV). The foundation of MFV is a Petrov-Galerkin primal-dual mixed formulation. More precisely, let us consider as model problem the following 2D-problem:

$$(1) \quad \begin{cases} div \ \mathbf{u} = f & \text{in } \Omega \\ \mathbf{u} = -\mathcal{A} \ \mathbf{grad} \ p & \text{in } \Omega \\ (\mathcal{A} \ \mathbf{grad} \ p).\mathbf{n} = 0 & \text{on } \partial\Omega, \end{cases}$$

where Ω is a connected open bounded domain of $\mathbb{R}^2$, $\mathbf{n}$ is the exterior unit normal vector along $\partial\Omega$, f is a function of $L^2(\Omega)$ and for all x in $\overline{\Omega}$, $\mathcal{A} = \mathcal{A}(x)$ is a 2×2 matrix. We also assume that f satisfies the compatibility condition $\int_\Omega f \, dx = 0$ and we assume that $\mathcal{A}$ is piecewise continuous on $\overline{\Omega}$ and that $\mathcal{A}(x)$ is a symmetric matrix which verifies a standard uniform ellipticity condition:

$$\exists \tau > 0, \ \forall \boldsymbol{\xi} \in \mathbb{R}^2, \ \boldsymbol{\xi}^T \mathcal{A}(x) \boldsymbol{\xi} \geq \tau \boldsymbol{\xi}^T \boldsymbol{\xi} \quad \text{for all } x \text{ in } \overline{\Omega}.$$

After elimination of the vectorial unknown $\mathbf{u}$, we see that the scalar unknown p is characterized as solution of the homogeneous Neumann problem:

$$(2) \quad \begin{cases} - div \ (\mathcal{A} \ \mathbf{grad} \ p) = f & \text{in } \Omega \\ (\mathcal{A} \ \mathbf{grad} \ p).\mathbf{n} = 0 & \text{on } \partial\Omega \end{cases}$$

In fact, p is defined up to a constant; we can fix this constant for our analysis by imposing the condition $\int_\Omega p \, dx = 0$.

Before giving the variational formulation of this problem, we introduce the functional framework: let $H^1(\Omega)$ be the Sobolev space

$$H^1(\Omega) = \left\{ p \in L^2(\Omega), \; \mathbf{grad}\, p \in (L^2(\Omega))^2 \right\}$$

and $H(div; \Omega)$ the space

$$H(div; \Omega) = \left\{ \mathbf{u} \in \left(L^2(\Omega) \right)^2 ; \; div\, \mathbf{u} \in L^2(\Omega) \right\}.$$

Then a primal-dual mixed formulation of (1) consists in seeking $(\mathbf{u}, p)$, solution of:

(3)
$$\begin{cases} (\mathbf{u}, p) \in \mathbf{U} \times P \\[2mm] \forall \mathbf{v} \in (L^2(\Omega))^2, \; \displaystyle\int_\Omega \mathbf{u}.\mathbf{v}\, dx + \int_\Omega \mathcal{A}\, \mathbf{grad}\, p.\mathbf{v}\, dx = 0 \\[2mm] \forall q \in L^2(\Omega), \; \displaystyle\int_\Omega q\, div\, \mathbf{u}\, dx = \int_\Omega fq\, dx, \end{cases}$$

with

(4)
$$\mathbf{U} = \left\{ \mathbf{u} \in \mathbf{H}(div, \Omega), \; \mathbf{u}.\mathbf{n} = 0 \; \text{ on } \partial\Omega \right\}$$
$$P = \left\{ p \in H^1(\Omega), \; \int_\Omega p\, dx = 0 \right\}.$$

Remark: In a classical dual mixed formulation, we seek $\mathbf{u}$ in $H(div; \Omega)$ and p in $L^2(\Omega)$; in primal mixed formulation we seek $\mathbf{u}$ in $(L^2(\Omega))^2$ and p in $H^1(\Omega)$. Here we have a primal-dual mixed formulation since $\mathbf{u}$ is an element of $H(div; \Omega)$ and p is an element of $H^1(\Omega)$. In (3), the test space of functions $(\mathbf{v}, q)$ is different of the trial space $\mathbf{U} \times P$: we have a Petrov-Galerkin formulation. Without elaborating any mixed theory, it is easy to verify directly that problem (3) has a solution and only one.

To the formulation (3), we associate a discrete formulations in the coupled unknowns $\mathbf{u}_h$ and p_h by giving us four finite dimensional subspaces of U, P, $(L^2(\Omega))^2$ and $L^2(\Omega)$. By using ad hoc numerical quadrature, we will obtain examples where the vectorial unknown $\mathbf{u}_h$ can be directly eliminated and then we will get a finite volume scheme in the scalar unknown p_h. In [16], we have given examples as well as a complete analysis of MFV with quadrangular finite elements for $\mathbf{u}_h$ and p_h on dual meshes; for examples of rectangular elements for $\mathbf{u}_h$ and triangular elements for p_h , see also [15]. In the same spirit but starting from the classical Bubnov-Galerkin dual mixed formulation: $\mathbf{u}$ and test functions $\mathbf{v}$ in $H(div; \Omega)$, p and test functions q in $L^2(\Omega)$, Baranger, Oudin and Maître ([1] and [2]) obtain MFV on triangular meshes but they have no possibility to extend their results to the general 3D-situation. In the contrary, it is straightforward to extend our examples to hexahedral elements on unstructured 3D-meshes.

2 Abstract theory of generalized mixed formulation

Let us consider the abstract problem:

$$(5) \qquad \begin{cases} (\mathbf{u}, p) \in \mathbf{U} \times P \\ \forall \mathbf{v} \in \mathbf{V}, \quad m(\mathbf{u}, \mathbf{v}) + a(p, \mathbf{v}) = 0 \\ \forall q \in Q, \quad b(\mathbf{u}, q) = \langle f, q \rangle_{Q', Q} \end{cases}$$

where $(\mathbf{U}, \|.\|_\mathbf{U})$, $(P, \|.\|_P)$, $(\mathbf{V}, \|.\|_\mathbf{V})$ and $(Q, \|.\|_Q)$ are Hilbert spaces; $m(.,.)$, $a(.,.)$ and $b(.,.)$ are continuous bilinear forms defined respectively on $\mathbf{U} \times \mathbf{V}$, $P \times \mathbf{V}$ and $\mathbf{U} \times Q$ and f is an element of Q', the topological dual space of Q. We assume that problem (5) has a unique solution $(\mathbf{u}, p)$; for sufficient conditions which imply, for any $f \in Q'$, the existence and the uniqueness of the solution of problem (5), see Nicolaïdes [10] and also [3].

Let $\mathbf{U}_h$, $\mathbf{V}_h$, P_h and Q_h be finite dimensional subspaces of $\mathbf{U}$, $\mathbf{V}$, P and Q respectively. We also consider $m_h(.,.)$ and $a_h(.,.)$ two bilinear forms defined on $\mathbf{U}_h \times \mathbf{V}_h$ and $P_h \times \mathbf{V}_h$ respectively. In the applications, $m_h(.,.)$ and $a_h(.,.)$ will be deduced from $m(.,.)$ and $a(.,.)$ by numerical quadratures. Notice that we don't try to approximate the bilinear form $b(.,.)$ which is used in the applications for expressing the conservation law. Next, we associate to problem (5) the discrete problem

$$(6) \qquad \begin{cases} (\mathbf{u}_h, p_h) \in \mathbf{U}_h \times P_h \\ \forall \mathbf{v}_h \in \mathbf{V}_h, \quad m_h(\mathbf{u}_h, \mathbf{v}_h) + a_h(p_h, \mathbf{v}_h) = 0 \\ \forall q_h \in Q_h, \quad b(\mathbf{u}_h, q_h) = \langle f, q_h \rangle_{Q', Q} \end{cases}$$

We can now state the following abstract result:

THEOREM 1. *Assume that*

$$(7) \qquad \dim(\mathbf{U}_h) + \dim(P_h) = \dim(\mathbf{V}_h) + \dim(Q_h)$$

and that the next three Babuska–Brezzi conditions are satisfied:

$$(8) \qquad i) \quad \inf_{\mathbf{u}_h \in \mathbf{U}_{0h} \backslash \{0\}} \sup_{\mathbf{v}_h \in \mathbf{V}_h} \frac{m_h(\mathbf{u}_h, \mathbf{v}_h)}{\|\mathbf{u}_h\|_\mathbf{U} \|\mathbf{v}_h\|_\mathbf{V}} \geq \mu > 0$$

$$(9) \qquad ii) \quad \inf_{p_h \in P_h \backslash \{0\}} \sup_{\mathbf{v}_h \in \mathbf{V}_{1h}} \frac{a_h(p_h, \mathbf{v}_h)}{\|p_h\|_P \|\mathbf{v}_h\|_\mathbf{V}} \geq \alpha > 0$$

$$(10) \qquad iii) \quad \inf_{q_h \in Q_h \backslash \{0\}} \sup_{\mathbf{u}_h \in \mathbf{U}_h} \frac{b(\mathbf{u}_h, q_h)}{\|\mathbf{u}_h\|_\mathbf{U} \|q_h\|_Q} \geq \beta > 0$$

where the subspaces $\mathbf{U}_{0h}$ and $\mathbf{V}_{1h}$ are defined by

$$\mathbf{U}_{0h} = \{ \mathbf{u}_h \in \mathbf{U}_h, \quad \forall q_h \in Q_h, \quad b(\mathbf{u}_h, q_h) = 0 \}$$

$$\mathbf{V}_{1h} = \{ \mathbf{v}_h \in \mathbf{V}_h, \quad \forall \mathbf{u}_h \in \mathbf{U}_{0h}, \quad m_h(\mathbf{u}_h, \mathbf{v}_h) = 0 \} .$$

165

Then Problem (6) has a unique solution $(\mathbf{u}_h, p_h)$.

THEOREM 2. *Assume the hypothesis (7) and (8)-(10) with α, μ and β independent of h; assume moreover that we can find constants M and A independent of h such that*

$$\forall \mathbf{u}_h \in \mathbf{U}_h, \quad \forall \mathbf{v}_h \in \mathbf{V}_h, \quad m_h(\mathbf{u}_h, \mathbf{v}_h) \leq M \, \|\mathbf{u}_h\|_{\mathbf{U}} \, \|\mathbf{v}_h\|_{\mathbf{V}}$$

$$\forall p_h \in P_h, \quad \forall \mathbf{v}_h \in \mathbf{V}_h, \quad a_h(p_h, \mathbf{v}_h) \leq A \, \|p_h\|_P \, \|\mathbf{v}_h\|_{\mathbf{V}}$$

Then there exists a constant C independent of h such that the following a priori error estimate holds:

$$\|p - p_h\|_P + \|\mathbf{u} - \mathbf{u}_h\|_{\mathbf{U}} \leq C \left\{ \inf_{r_h \in P_h} \left(\|p - r_h\|_P + \sup_{\mathbf{v}_h \in \mathbf{V}_h} \frac{|a(r_h, \mathbf{v}_h) - a_h(r_h, \mathbf{v}_h)|}{\|\mathbf{v}_h\|_{\mathbf{V}}} \right) \right.$$

(11)

$$\left. + \inf_{\mathbf{w}_h \in \mathbf{U}_h} \left(\|\mathbf{u} - \mathbf{w}_h\|_{\mathbf{U}} + \sup_{\mathbf{v}_h \in \mathbf{V}_h} \frac{|m(\mathbf{w}_h, \mathbf{v}_h) - m_h(\mathbf{w}_h, \mathbf{v}_h)|}{\|\mathbf{v}_h\|_{\mathbf{V}}} \right) \right\}.$$

For a detailed of a proof of Theorem 1 and Theorem 2, see [17]. With similar, but lightly different Babuška-Brezzi conditions, analogous results are due to Nicolaïdes [10], see also [3].

In the next section, we shall use this result to analyze the discretization deduced from the formulation (3). For that, we have to build finite dimensional spaces which verify all the conditions of Theorems 1 and 2.

3 Triangular Finite Volume Methods

3.1 Triangulations

Let $(\mathcal{T}_h)_h$ be a regular family of triangulations of $\bar{\Omega}$, $\mathcal{T}_h$ being composed only of triangles T_l, $1 \leq l \leq L$, with diameters $\leq h$. We denote by G_l the center of gravity (i.e. the intersection of medians) of a triangle T_l.

By connecting for $l = 1, ..., L$, the center G_l to the three vertices of T_l and to the three edge midpoints of T_l, we build a subtriangulation $\mathcal{T}_h^{\#}$ of $\bar{\Omega}$. The triangles of $\mathcal{T}_h^{\#}$ will be denoted $T_m^{\#}$, $1 \leq m \leq M$. The family $(\mathcal{T}_h^{\#})_h$ is also a regular family of triangulations of $\bar{\Omega}$. One can remark that the primal triangulation $\mathcal{T}_h$ has L triangles while the subtriangulation $\mathcal{T}_h^{\#}$ has $M = 6\,L$ triangles.

Let I be the number of vertices of the primal triangulation $\mathcal{T}_h$. To any vertex $S_i, 1 \leq i \leq I$, of $\mathcal{T}_h$, we associate a control volume K_i defined as the union of all the triangles of $\mathcal{T}_h^{\#}$ which admit S_i as vertex. We obtain in this manner a new triangulation $\mathcal{T}_h^*$ which is called the dual one of $\mathcal{T}_h$ $\bar{\Omega} = \underset{1 \leq i \leq I}{\cup} K_i$; of course, the cells K_i of the triangulation $\mathcal{T}_h^*$ are polygonal, not necessarily convex. This dual triangulation has I cells.

166

Let $J^{\#}$ be the number of edges of the subtriangulation $\mathcal{T}_h^{\#}$ which are not situated on the boundary $\partial\Omega$. Let $\Gamma_j^{\#}$, $1 \leq j \leq J^{\#}$, be such an edge: $\Gamma_j^{\#}$ is the common edge of two triangles $T_m^{\#}$ and $T_n^{\#}$ of $\mathcal{T}_h^{\#}$; we will denote $K_j^{\#}$ the set $K_j^{\#} = T_m^{\#} \cup T_n^{\#}$. The unit vector $\mathbf{n}_{m,n}^{\#}$ normal to $T_m^{\#} \cap T_n^{\#} = \Gamma_j^{\#}$ is directed from $T_m^{\#}$ towards $T_n^{\#}$, so one has $\mathbf{n}_{m,n}^{\#} = -\mathbf{n}_{n,m}^{\#}$; let $\mathbf{n}_j^{\#}$ be an arbitrary choice between $\mathbf{n}_{m,n}^{\#}$ or $\mathbf{n}_{n,m}^{\#}$. For any j, $1 \leq j \leq J^{\#}$, we will introduce the piecewise constant vectorial function $\mathbf{w}_j^{\#}$ defined by

$$\mathbf{w}_j^{\#} = \left\{ \begin{array}{ll} \mathbf{n}_j^{\#} & \text{on } K_j^{\#} \\ \mathbf{0} & \text{on } \bar{\Omega} - K_j^{\#}. \end{array} \right.$$

3.2 Discretization spaces

We can now introduce the discretization spaces.
First we choose the space associated with the scalar unknown p_h:

$$(12) \qquad P_h = \{ p_h \in P; \ \forall T_l \in \mathcal{T}_h, \ p_h |_{T_l} \in P_1(T_l) \},$$

where $P_1(T_l)$ is the space of affine functions on T_l.
Then we take as space of the vectorial unknown $\mathbf{u}_h$:

$$(13) \qquad \mathbf{U}_h = \left\{ \mathbf{u}_h \in \mathbf{U}; \ \forall T_m^{\#} \in \mathcal{T}_h^{\#}, \ \mathbf{u}_h |_{T_m^{\#}} \in \mathbf{D}_1(T_m^{\#}) \right\}$$

where $\mathbf{D}_1(T_m^{\#})$ is the space constituted by the functions of $\left(P_1(T_m^{\#}) \right)^2$ whose normal traces along each edge of $T_m^{\#}$ are constant; this space is known as the lowest order Raviart-Thomas space.
We now introduce the spaces associated with the test functions. First we choose for vectorial test functions the following space

$$(14) \qquad \mathbf{V}_h = \text{Span} \left\{ \mathbf{w}_j^{\#}, \ 1 \leq j \leq J^{\#}, \right\}$$

and finally we take for scalar test functions the space

$$(15) \qquad Q_h = \left\{ q_h \in L^2(\Omega) \text{ with } \int_\Omega q_h \, dx = 0; \ \forall K_i \in T_h^{*}, \ q_h |_{T_i^{*}} \in P_0(K_i) \right\},$$

where $P_0(K_i)$ is the space of constant functions on K_i.

It is clear that with these choices of finite dimensional spaces we have

$$\left\{ \begin{array}{l} \dim(P_h) = \dim(Q_h) = I - 1 \\ \dim(\mathbf{U}_h) = \dim(\mathbf{V}_h) = J^{\#}, \end{array} \right.$$

so the assumption (7) is verified.

3.3 Approximated bilinear forms

Let us begin by defining the bilinear form $m(.,.)$ on $H\left(div;\Omega\right)\times(L^2(\Omega))^2$ by

$$m(\mathbf{u},\mathbf{v}) = \int_{\Omega} \mathbf{u}.\mathbf{v}\, dx.$$

In order to obtain a volume finite method, we define the bilinear form $m_h(.,.)$ on $\mathbf{U}_h \times \mathbf{V}_h$ by

$$(16) \qquad \forall \mathbf{u}_h \in \mathbf{U}_h,\ \forall 1 \le j \le J^{\#},\quad m_h(\mathbf{u}_h, \mathbf{w}_j^{\#}) = \mathrm{meas}(K_j^{\#})\,(\mathbf{u}_h.\mathbf{n}_j^{\#})\,|_{\Gamma_j^{\#}}$$

One recalls that for $\mathbf{u}_h$ in $\mathbf{U}_h$, $\mathbf{u}_h.\mathbf{n}_j^{\#}$ is constant on $\Gamma_j^{\#}$. Since the quantity $m(\mathbf{u}_h, \mathbf{w}_j^{\#})$ is an integral over the domain $K_j^{\#}$, $m_h(\mathbf{u}_h, \mathbf{w}_j^{\#})$ can be considered as an approximation of $m(\mathbf{u}_h, \mathbf{w}_j^{\#})$ by a one point numerical quadrature rule over $K_j^{\#}$.

The bilinear form $a(.,.)$ on $H^1(\Omega) \times (L^2(\Omega))^2$ is next defined by

$$a(p,\mathbf{v}) = \int_{\Omega} \mathcal{A}\,\mathbf{grad}\,p.\mathbf{v}\,dx$$

and this last expression is replaced in the discrete problem by

$$(17) \qquad \forall p_h \in P_h,\ \forall \mathbf{v}_h \in \mathbf{V}_h,\quad a_h(p_h, \mathbf{v}_h) = \int_{\Omega} \mathcal{A}_h\,\mathbf{grad}\,p_h.\mathbf{v}_h\,dx,$$

where $\mathcal{A}_h$ is the matrix constant on each triangle $T_l \in \mathcal{T}_h$ such that

$$\mathcal{A}_h\,|_{T_l} = \mathcal{A}_h\,(G_l)\,.$$

Finally, we introduce the bilinear form $b(.,.)$ defined on $H\left(div,\Omega\right)\times L^2(\Omega)$ by

$$b(\mathbf{u},q) = \int_{\Omega} q\, div\,\mathbf{u}\,dx,$$

which expresses the conservation law. Therefore, in order to obtain a conservative scheme, it is essential to use none numerical integration on these quantities; we have

$$(18) \qquad \forall \mathbf{u}_h \in \mathbf{U}_h,\ \forall q_h \in Q_h,\quad b(\mathbf{u}_h, q_h) = \int_{\Omega} q_h\, div\,\mathbf{u}_h\,dx.$$

3.4 Discrete problem

With all these notations $(12)-(19)$, the discrete problem can now be written as below:

$$(19) \qquad \begin{cases} (\mathbf{u}_h, p_h) \in \mathbf{U}_h \times P_h \\ \forall \mathbf{v}_h \in \mathbf{V}_h,\ m_h(\mathbf{u}_h, \mathbf{v}_h) + a_h(p_h, \mathbf{v}_h) = 0 \\ \forall q_h \in Q_h,\ b(\mathbf{u}_h, q_h) = \langle f, q_h \rangle_{Q',Q} \end{cases}$$

Before any other consideration, let us notice that, since $\int_\Omega div\, \mathbf{u}_h\, dx = \int_\Omega f\, dx = 0$, the last row of (19) translates into

$$\forall K_i \in T_h^*, \quad \int_{K_i} div\, \mathbf{u}_h\, dx = \int_{K_i} f\, dx,$$

or, equivalently,

$$(20) \qquad \forall K_i \in T_h^*, \quad \int_{\partial K_i} \mathbf{u}_h.\mathbf{n}_{\partial K_i}\, d\sigma = \int_{K_i} f\, dx$$

where $\mathbf{n}_{\partial K_i}$ is the unit outward normal along ∂K_i.

Next, thanks to the choice of the test functions $\mathbf{V}_h$ and of the approximated form $m_h(.,.)$, one can remark that the second row of (19) allows an explicit elimination of the degrees of freedom of $\mathbf{u}_h$. This leads to a scheme in only the degrees of freedom of p_h, i.e. the values of p_h at the vertices of the triangulations if we forgot the global constraint $\int_\Omega p_h\, dx = 0$.

THEOREM 3. *Problem (19) has a unique solution* $(p_h, \mathbf{u}_h)$.

Proof: Once the inf-sup conditions (8)-(10) established in the present concrete situation, Theorem 3 follows as a corollary of the abstract Theorem 1. These inf-sup conditions will be necessary for obtaining a priori error estimates. We give hereafter a direct proof of Theorem 3. By linearity, thanks to (7), it is sufficient to prove that the unique solution $(p_h, \mathbf{u}_h) \in \mathbf{U}_h \times P_h$ of

$$\begin{cases} \forall \mathbf{v}_h \in \mathbf{V}_h, \quad m_h(\mathbf{u}_h, \mathbf{v}_h) + a_h(p_h, \mathbf{v}_h) = 0 \\ \forall q_h \in Q_h, \quad b(\mathbf{u}_h, q_h) = 0 \end{cases}$$

is the trivial solution. The second equation is equivalent to

$$\forall 1 \le i \le I, \quad \int_{K_i} div\, \mathbf{u}_h\, dx = 0,$$

which we can also write:

$$(21) \qquad \forall 1 \le i \le I, \quad \int_{\partial K_i} \mathbf{u}_h.\mathbf{n}_{\partial K_i}\, d\sigma = 0.$$

The first equation is equivalent to

$$(22) \quad \forall 1 \le j \le J^\#, \quad meas\left(K_j^\#\right)\left(\mathbf{u}_h.n_j^\#\right)\big|_{\Gamma_j^\#} + \int_{K_j^\#} A_h\, \mathbf{grad}\, p_h.\mathbf{w}_j^\#\, dx = 0.$$

Since the quantities $\left(\mathbf{u}_h . n_j^{\#}\right)\big|_{\Gamma_j^{\#}}$ are the degrees of freedom of $\mathbf{u}_h$, in order to prove Theorem 3 it remains to show that $(21) - (22)$ imply $p_h = 0$.

For any i, $1 \leq i \leq I$, let $\Gamma_j^{\#}$ be an edge of ∂K_i : by construction of the triangulations, $\Gamma_j^{\#}$ is the median of some triangular cell $K_j^{\#}$ and this triangle $K_j^{\#}$ is included in some triangle $T_l \in \mathcal{T}_h$ which admits S_i as vertex. The triangle $K_j^{\#}$ is the union of two triangles $T_m^{\#}$, $T_n^{\#} \in \mathcal{T}_h^{\#}$; the triangle of $\mathcal{T}_h^{\#}$ which has S_i as vertex and $\Gamma_j^{\#}$ as opposite edge will be denoted $T_{i,j}^{\#}$: so one has $T_{i,j}^{\#} = T_m^{\#}$, or otherwise $T_{i,j}^{\#} = T_n^{\#}$.

With the previous notations, since $K_j^{\#} \subset T_l$ the function $\mathbf{grad}\, p_h . \mathbf{w}_j^{\#}$ is constant on $K_j^{\#}$; moreover by construction the approximated matrix $\mathcal{A}_h$ is constant on T_l : $\mathcal{A}_h(x) = \mathcal{A}_h(G_l)$ for all $x \in T_l$. So for any edge $\Gamma_j^{\#}$ of ∂K_i, the above relation can be written as:

$$\left(\mathbf{u}_h + \mathcal{A}_h \, \mathbf{grad}\, p_h\right).n_j^{\#} = 0 \ \text{ on } \Gamma_j^{\#}$$

By summation over j such that $\Gamma_j^{\#}$ is an edge of ∂K_i, we obtain

$$\int_{\partial K_i} \left(\mathbf{u}_h + \mathcal{A}_h \, \mathbf{grad}\, p_h\right).\mathbf{n}_{\partial K_i}\, d\sigma = 0$$

and therefore by the means of (21):

$$(23) \qquad \int_{\partial K_i} \left(\mathcal{A}_h \, \mathbf{grad}\, p_h\right).\mathbf{n}_{\partial K_i}\, d\sigma = 0$$

The proof of Theorem 3 is now a consequence of the following lemma. In the sequel, the same notation: $\|\,.\,\|_{0,\Omega}$ is used for the norm of $L^2(\Omega)$ or of $\left(L^2(\Omega)\right)^2$.

LEMMA 1. *There exists a constant $a > 0$ such that the inequality*

$$(24) \qquad -\sum_{1 \leq i \leq I} \left\{ p_h(S_i) \int_{\partial K_i} \left(\mathcal{A}_h \, \mathbf{grad}\, p_h\right).\mathbf{n}_{\partial K_i}\, d\sigma \right\} \geq a \, \|\mathbf{grad}\, p_h\|_{0,\Omega}^2$$

holds for all $p_h \in P_h$.

Proof: Let $X_l = -\sum_{1 \leq i \leq I} \left\{ p_h(S_i) \int_{\partial K_i \cap T_l} \left(\mathcal{A}_h \, \mathbf{grad}\, p_h\right).\mathbf{n}_{\partial K_i}\, d\sigma \right\}$ be the contribution due to a triangle $T_l \in \mathcal{T}_h$. Since $\mathcal{A}_h$ and $\mathbf{grad}\, p_h$ are constant on T_l, we have

$$X_l = -\left(\mathcal{A}_h \, \mathbf{grad}\, p_h\right)\big|_{T_l} \cdot \sum_{1 \leq i \leq I} \left\{ \int_{\partial K_i \cap T_l} p_h(S_i)\, \mathbf{n}_{\partial K_i}\, d\sigma \right\}$$

Assume for a moment that T_l is equilateral of diameter h. Then an elementary computation gives us:

$$-\sum_{1 \leq i \leq I} \left\{ \int_{\partial K_i \cap T_l} p_h(S_i)\, \mathbf{n}_{\partial K_i}\, d\sigma \right\} = \frac{3}{2}\, meas(T_l)\, (\mathbf{grad}\, p_h)\big|_{T_l}$$

170

and so it comes in this particular case:

$$X_l = \tfrac{3}{2} \, meas\,(T_l) \; (\mathcal{A}_h \, \mathbf{grad} \, p_h)\,|_{T_l} \cdot (\mathbf{grad} \, p_h)\,|_{T_l}$$
$$\geq a \, \| \, \mathbf{grad} \, p_h \|_{0,T_l}^2 \; .$$

We treat now the general case. First we transform by an affine mapping the triangle T_l into an equilateral triangle $\widehat{T}$; if DF_l is the Jacobian matrix of the transformation F_l from $\widehat{T}$ onto T_l and $(DF_l)^{-T}$ the transposed inverse matrix, then we have

$$X_l = - \left(\widehat{\mathcal{A}_h} \, \mathbf{grad} \, \widehat{p_h} \right) |_{\widehat{T}} \cdot \sum_{1 \leq i \leq I} \left\{ \int_{\widehat{\partial K_i \cap T_l}} \widehat{p_h} \left(\widehat{S_i} \right) \mathbf{n}_{\widehat{\partial K_i \cap T_l}} \, d\sigma \right\}$$

with

$$\widehat{\mathcal{A}_h} = |\det (DF_l)| \, (DF_l)^{-1} \, \mathcal{A}_h \, (DF_l)^{-T}$$

Then we deduce

$$X_l \geq a_l \, \| \, \mathbf{grad} \, \widehat{p_h} \|_{0,\widehat{T}}^2$$

and coming back to the triangle T_l :

$$X_l \geq a \, \| \, \mathbf{grad} \, p_h \|_{0,T_l}^2$$

for some constant $a > 0$. By summation over l, $1 \leq l \leq L$, the Lemma is proved. Note that we can choose the constant a independent of h, which will be useful for the error estimates.

4 A priori error estimates

We will use in this Section the abstract result given by the Theorem 2. For that, we introduce the Hilbert space norms :

$$\| \, \mathbf{u} \, \|_{\mathbf{U}} = \| \, \mathbf{u} \, \|_{H(div,\Omega)} = \left\{ \| \, \mathbf{u} \, \|_{0,\Omega}^2 + \| \, div \, \mathbf{u} \, \|_{0,\Omega}^2 \right\}^{1/2}$$

and

$$\| \, p \, \|_P = \| \, p \, \|_{H^1(\Omega)} = \left\{ \| \, p \, \|_{0,\Omega}^2 + \| \, \mathbf{grad} \, p \, \|_{0,\Omega}^2 \right\}^{1/2}$$

The spaces Q and V are equipped with the norms $\| \, . \, \|_{0,\Omega}$ of $L^2\,(\Omega)$ and $(L^2\,(\Omega))^2$ respectively.

4.1 Inf-sup properties

It is easy to prove the $\inf - \sup$ condition (8) with a constant $\mu > 0$ independent of h. The $\inf - \sup$ condition (10) with a constant $\beta > 0$ independent of h results immediately from a classical $\inf - \sup$ condition for the mixed Raviart-Thomas finite element theory, see [12]. The inf − sup condition (9) with a constant $\alpha > 0$ independent of h is more delicate. To any function $p_h \in P_h$ we associate the vectorial function $\mathbf{v}_h \in \mathbf{V}_h$ defined by $\overline{\mathbf{v}_h} = \sum_{1 \leq j \leq J\#} \nu_j \, \mathbf{w}_j^{\#}$ with:

$$
\begin{cases}
\nu_j = \dfrac{meas\left(\Gamma_j^{\#}\right)}{2\,meas\left(K_j^{\#}\right)} \left\{p_h\left(S_{i_1}\right) - p_h\left(S_{i_2}\right)\right\} & \text{if } \Gamma_j^{\#} = \partial K_{i_1} \cap \partial K_{i_2} \text{ and } \mathbf{w}_j^{\#} = n_{\partial K_{i_1}} \\[2mm]
\nu_j = \dfrac{meas\left(\Gamma_j^{\#}\right)}{2\,meas\left(K_j^{\#}\right)} \left\{p_h\left(S_{i_2}\right) - p_h\left(S_{i_1}\right)\right\} & \text{if } \Gamma_j^{\#} = \partial K_{i_1} \cap \partial K_{i_2} \text{ and } \mathbf{w}_j^{\#} = n_{\partial K_{i_2}} \\[2mm]
\nu_j = 0 & \text{if } \Gamma_j^{\#} \text{ is not included in } \cup_{1 \leq i \leq I} \, \partial K_i.
\end{cases}
$$

For this choice, we have on the one hand

$$
a_h\left(p_h, \overline{\mathbf{v}_h}\right) = -\left(\mathcal{A}_h \, \mathbf{grad}\, p_h\right) |_{T_l} \cdot \sum_{1 \leq i \leq I} \left\{\int_{\partial K_i \cap T_l} p_h\left(S_i\right) \, \mathbf{n}_{\partial K_i} \, d\sigma\right\}
$$

and apply Lemma 1, there exists $a > 0$ such that

$$
a_h\left(p_h, \overline{\mathbf{v}_h}\right) \geq a \, \|\mathbf{grad}\, p_h\|_{0,\Omega}^2 \,.
$$

On the other hand, one can establish that

$$
\|\overline{\mathbf{v}_h}\|_{0,\Omega} \leq c \, \|\mathbf{grad}\, p_h\|_{0,\Omega}
$$

We can now deduce from the last two relations the inf − sup condition (9) with a constant $\alpha > 0$ independent of h.

4.2 Stability and approximation properties

The proofs of the following lemmas are without any particular difficulty:

LEMMA 2. *There exists a constant M independent of h such that*

$$
\forall \mathbf{u}_h \in \mathbf{U}_h, \quad \forall \mathbf{v}_h \in \mathbf{V}_h, \quad m_h(\mathbf{u}_h, \mathbf{v}_h) \leq M \, \|\mathbf{u}_h\|_{0,\Omega} \, \|\mathbf{v}_h\|_{0,\Omega}
$$

LEMMA 3. *There exists a constant A independent of h such that*

$$
\forall p_h \in P_h, \quad \forall \mathbf{v}_h \in \mathbf{V}_h, \quad a_h(p_h, \mathbf{v}_h) \leq A \, \|p_h\|_{1,\Omega} \, \|\mathbf{v}_h\|_{0,\Omega}
$$

LEMMA 4. *Assume that $p \in H^2(\Omega)$ and that the coefficients of $\mathcal{A}$ are in $C^1(\overline{\Omega})$. Then there exists a constant C independent of h such that*

$$
\inf_{r_h \in P_h} \left(\|p - r_h\|_{1,\Omega} + \sup_{\mathbf{v}_h \in \mathbf{V}_h} \frac{|a(r_h, \mathbf{v}_h) - a_h(r_h, \mathbf{v}_h)|}{\|\mathbf{v}_h\|_{0,\Omega}}\right) \leq C h.
$$

LEMMA 5. *Assume that* $\mathbf{u} \in (H^1(\Omega))^2$ *and that div* $\mathbf{u} \in H^1(\Omega)$. *There exists a constant* C *independent of* h *such that*

$$\inf_{\mathbf{w}_h \in \mathbf{U}_h} (\|\mathbf{u} - \mathbf{w}_h\|_{\mathbf{H}(div,\Omega)} + \sup_{\mathbf{v}_h \in \mathbf{V}_h} \frac{|m(\mathbf{w}_h, \mathbf{v}_h) - m_h(\mathbf{w}_h, \mathbf{v}_h)|}{\|\mathbf{v}_h\|_{0,\Omega}}) \leq Ch.$$

4.3 Error bound

As conclusion, we now state the result which implies the convergence of the MFV:

THEOREM 4. *Assume the coefficients of the matrix* A *continuously differentiable on* $\bar{\Omega}$. *If the solution* $(\mathbf{u}, p)$ *of Problem (3) is sufficiently regular and if* $(\mathbf{u}_h, p_h)$ *is the solution of Problem (19) obtained with the choices* $(12) - (18)$, *then we have*

(25) $$\|\mathbf{u}_h - \mathbf{u}\|_{\mathbf{H}(div,\Omega)} + \|p - p_h\|_{1,\Omega} \leq Ch(|\mathbf{u}|_{1,\Omega} + |div\ \mathbf{u}|_{1,\Omega} + |p|_{2,\Omega})$$

References

[1] J. BARANGER, J.-F. MAÎTRE AND F. OUDIN (1994): *"Application de la théorie des éléments finis mixtes à l'étude d'une classe de schémas aux volumes différences finies pour les problèmes elliptiques"*, C. R. Acad. Sci. Paris, **319**, Série I, 401-404.

[2] J. BARANGER, J.-F. MAÎTRE AND F. OUDIN (1996): *"Connection between finite volume and mixed finite element methods"*, M2AN, **30**, 445-465.

[3] C. BERNARDI, C. CANUTO AND Y. MADAY (1988): *" Generalized Inf-Sup Conditions for Chebychev Spectral Approximation of Stokes Problem"*, SIAM J. Numer. Anal., **25**, 1237–1271.

[4] Z. CAI (1991): *"On the Finite Volume Element Method"*, Numer. Math., **58**, 713–735.

[5] Z. CAI, J. MANDEL AND S. MC CORMICK (1991): *"The Finite Volume Element Method for Diffusion Equations on General Triangulations"*, SIAM J. Numer. Anal., **28**, 392–402.

[6] R. EYMARD, T. GALLOUËT AND R. HERBIN : *"The Finite Volume Method"*, in preparation for the Handboof of Numerical Analysis", Ph. Ciarlet and J. L. Lions eds.

[7] I. FAILLE (1992): *"A Control Volume Method to Solve an Elliptic Equation on a Two Dimensional Irregular Mesh"*, Computer Methods in Applied Mechanics and Engineering, **100**, 275–290.

[8] T. GALLOUËT (1992): *"Cours sur les méthodes volumes finis"*, E.D.F., Clamart.

[9] B. HEINRICH (1987): "Finite Difference Methods on Irregular Networks, a Generalized Approach to Second Order Elliptic Problem", Birkhaüser Verlag.

[10] R. A. NICOLAÏDES (1982): *"Existence, Uniqueness and Approximation for Generalized Saddle–Point Problems"*, SIAM J. Numer. Anal., **19**, 349–357.

[11] P. A. RAVIART AND J.-M. THOMAS (1977): *" A Mixed Finite Element Method for Second Order Elliptic Problems"*, in Mathematical Aspects of the Finite Element Method, Lecture Notes in Mathematics, **606**, Springer–Verlag, 292–315.

[12] J. E. ROBERTS AND J.-M. THOMAS (1991): *"Mixed and Hybrid Methods"*, in Handbook of Numerical Analysis, Finite Element Methods, (Vol II), Finite Element Methods (Part 1), Ciarlet P.G. and Lions J.L. eds, North–Holland, Amsterdam, 523–639.

[13] T. RUSSELL, J. JONES, Z. CAI AND S. MCCORMICK (1995): *"A Control-Volume Mixed Method on Irregular Quadrilateral and Hexalateral Grids"*, to appear.

[14] E. SÜLI (1991):*"Convergence of the Finite Volume Schemes for Poisson's Equation on Nonuniform Meshes"*, SIAM J. Numer. Anal., **28**, 1419–1430.

[15] J.-M. THOMAS AND D. TRUJILLO (1994): *"Finite Volume Variational Formulation. Application to Domain Decomposition Methods"*, in Domain Decomposition Methods in Sciences and Engineering, Sixth International Conference on Domain Decomposition Methods in Science and Engineering (SIAM 15-19 June 1992, Como, Italy), A. Quarteroni et al editors, AMS, Series "Contemporary Mathematics", **157**, 127–132.

[16] J.-M. THOMAS AND D. TRUJILLO (1995): *"Analysis of finite Volume Method and Application to Reservoir Simulation"*, in Proceedings of the Conference Mathematical Modelling of Flow through Porous Media, (A. Bourgeat, C. Carasso S. Luckhaus, A. Mikelic eds), World Scientific Publishing, 318-396.

[17] D. TRUJILLO (1994): *"Couplage espace-temps de schémas numériques en simulation de réservoir"*, Thèse, Université de Pau et des Pays de l'Adour.